T0326469

Diffusion of Renewable Energy Technologies

Finanzmärkte und Klimawandel

Herausgegeben von
Dirk Schiereck und Paschen von Flotow

Band 4

Christian Friebe

Diffusion of
Renewable Energy Technologies
Private Sector Perspectives on Emerging Markets

Bibliographic Information published by the Deutsche Nationalbibliothek
The Deutsche Nationalbibliothek lists this publication
in the Deutsche Nationalbibliografie; detailed bibliographic
data is available in the internet at http://dnb.d-nb.de.

Zugl.: EBS Universität für Wirtschaft und Recht, Diss., 2013

Library of Congress Cataloging-in-Publication Data

Friebe, Christian, 1983-
Diffusion of renewable energy technologies : private sector perspectives on
emerging markets / Christian Friebe. -- 1 Edition.
   pages cm. -- (Finanzmärkte und Klimawandel ; Band 4)
Includes bibliographical references.
ISBN 978-3-631-64444-7
1. Renewable natural resources. 2. Renewable natural resources--Government
policy. 3. Solar energy--Government policy. 4. Wind power--Government policy.
I. Title.
HC85.F75 2014
333.79'4--dc23
                          2013049719

D 1540
ISSN 2190-3069
ISBN 978-3-631-64444-7 (Print)
E-ISBN 978-3-653-03111-9 (E-Book)
DOI 10.3726/ 978-3-653-03111-9

© Peter Lang GmbH
Internationaler Verlag der Wissenschaften
Frankfurt am Main 2014
All rights reserved.
PL Academic Research is an Imprint of Peter Lang GmbH.

Peter Lang – Frankfurt am Main · Bern · Bruxelles · New York ·
Oxford · Warszawa · Wien

This book is part of an editor's series of PL Academic Research
and was peer reviewed prior to publication.

www.peterlang.com

Whether we care most about national security, climate change, or jobs and competitiveness, we should do exactly the same things about energy.

Amory Lovins, 2010

Contents

Preface

In the global context of climate change and increasing energy demand, new methods of power generation and distribution are required. Renewable energy technologies could address these upcoming challenges. Moreover, they are currently becoming increasingly cost effective when compared to "conventional" fossil fuel based technologies for power generation.

The research papers specifically focus on mature renewable energy technologies in nascent markets in emerging and developing countries. Many of these countries face the challenge of delivering basic electricity services to a growing population. Worldwide, 1,5bn people do not have access to electricity services while an additional 1bn have only very limited access, although they are connected to the grid. In addition, many emerging and developing countries currently grapple with a huge increase in energy demand resulting from rapid economic development. The urgent need to overcome these challenges presents a vital opportunity to increase the share of renewable energy technologies, either in centralized "grid-connected" power systems or in decentralized "off-grid" power systems. Therefore, this set of research papers investigates a technology case study for each category, namely large-scale wind farms (grid-connected) and small-scale solar home systems (off-grid), through qualitative and quantitative methods. These two cases are further used to explore barriers of diffusion for off-grid and grid connected renewable energy technologies in general. For this purpose, innovative methods based on marketing research in consumer preferences have been selected and successfully applied. These methods are adapted to identify preferences for investment framework conditions for wind farms on one hand, and innovative business models for rural electrification on the other.

The key findings of the three research papers indicate that both technology applications face barriers which are relevant but conquerable. Some of the barriers refer to the institutional context, such as non-defined or not well adjusted regulations, uncertainties with respect to grid infrastructure developments and others. Other barriers are business- or private sector-related, such as the design of adequate products and services which address the needs of the rural population. Last but not least, several barriers result from inadequate funding mechanisms. It is concluded that, depending on the context of application, different risk and return structures of renewable energy applications require well-targeted public measures to increase the amount of private investments, thus enhancing diffusion.

By analyzing the perspective of private companies that are commercially successful early adopters of renewable energy applications, this work derives some first conclusions regarding effective and efficient public policy mechanisms. Thereby, this study also contributes to the debate connecting technology diffusion and public private partnerships (PPPs) on the national or international level. The cases studied here can be used as examples which inform the current debate on such PPP funds, like the Green Climate Fund currently being designed by the international community. On an applied level, it shows how business models for companies can be adjusted to specific market conditions in emerging and developing countries, how international value chains interact, and how companies and projects might be financed in highly regulated electricity markets. Hence, this study elaborates on a discussion agenda that seems imperative for a much-needed and concrete dialogue among public policy makers and private companies and investors at the national and international levels.

August 2013

Ronald Gleich, Strascheg Institute for Innovation and Entrepreneurship (SIIE)
Paschen von Flotow, Sustainable Business Institute (SBI)
Dirk Schiereck, Technical University Darmstadt

Acknowledgements

This thesis has been conducted as part of the research project "CFI - Climate Change, Financial Markets and Innovation" which is funded by the German Ministry for Research and Education (BMBF). Thereby, all three research papers benefited tremendously from the genuine support of Paschen von Flotow, head of the CFI research project and executive director of the Sustainable Business Institute (SBI). I would also like to thank Florian Täube, professor at the EBS University, who contributed significantly to the success of this thesis. Having two supervisors and co-authors like Paschen von Flotow and Florian Täube as well as the ongoing friendly and constructive dialogue with Ronald Gleich, head of the Strascheg Institute and professor at the EBS University, is truly the key success factor of this thesis.

During the process of research many people spend their valuable time by sharing and discussing their perspective with a young researcher like me. I am very grateful for experiencing the friendly support of many experts and decision makers during formal and informal interviews and both workshops. Also, it was a fantastic experience to expose intermediate results of the thesis at conferences and workshops. Beyond the formal and informal exchange with other researchers, two out of the three research papers benefited from an academic research grant given by "Sawtooth Software", who allowed me to use their software tools. I gratefully acknowledge their support.

Doing research would not be so much fun without the great teams of coworkers and friends at the SIIE and the SBI. I want to thank Anke, Conny, Daniel, Daniela, Dennis, Florian, Friedemann, Julian, Marco, Michael and Ruhi for the great time we had during and after work. I would also like to acknowledge many of my friends who supported this thesis by helping me to either focus on doing better research or occasionally urge me to relax and set it aside.

I am forever indebted to my mother Inge and my father Rainer as well as his wife Regina, who tirelessly encouraged me to develop my own ideas and follow my heart. Finally, I am especially grateful for my wonderful wife Sophie who was infinitely patient when my mind was once again "lost in research".

November 2013

Christian Friebe

List of Figures

List of Tables

1 Introduction

1.1 Purpose and scope

In many countries, the energy system is experiencing a significant transition from fossil fuels towards renewable energy. The key reasons for this transition are pressing global issues such as increasing and more volatile fossil fuel prices or environmental considerations such as climate change that potentially impacts all sectors of society (IEA, 2008; Huberty and Zysman, 2010). Emerging and developing countries face the additional challenge of high increase in energy demand, but might benefit of leapfrogging from fossil fuel based energy supply directly to the use of mature renewable energy technologies (Soete, 1985; Chircu and Mahajan, 2009).

As a reliable energy supply is imperative for economic development, the energy market in most countries is heavily regulated by policy-makers. This is especially true for the electricity sector, which requires extensive infrastructure in order to deliver electricity to the end-user. Depending on regulations as well as other contextual factors, long-term investments in large centralised and small decentralised power plants can be very attractive to private sector decision-makers. While there is widespread agreement that favourable policies can facilitate adoption and diffusion, there is no consensus about what exactly "favourable" means. Therefore, policy-makers often develop and adapt regulations without approved evidence for the perception of and the potential effects on private companies and investors (Wiser and Pickle, 1998). The aim of the thesis is to shed light on the perspective of private companies that are focussing on renewable energy technologies in the context of emerging and developing countries. Hence, this thesis seeks to contribute to the following debate:

What can policy-makers learn from the private sector perspective on renewable energy diffusion in the context of emerging and developing countries?

In order to do so, two specific renewable energy technologies, namely wind farms and solar home systems, are analysed from a private sector perspective. Both technologies are comparatively mature which allows for abstraction from technological risks and to focus on remaining influencing factors for adoption and diffusion. Thereby the thesis aims to help policy-makers as well as public investors to understand the private sector's point of view and to consequently make more informed decisions.

1.2 Diffusion of innovation

Rogers (1995) defines diffusion as "the process by which an innovation is communicated through certain channels over time among members of a social system". The 'innovation' mentioned may refer to a process, idea, concept or product, or a combination of these, which is freshly available to potential adopters. In this thesis, the 'innovation' refers to energy technology which is new not because of technological innovation, but on account of deployment in a new region with high market potential, but little or no legacy of the technology's application. In order to understand influencing factors for technology adoption within the social system (Figure 1.1) or more specifically, within the institutional framework (Porter, 1990; Wong, 2005), the thesis focuses firstly, on innovators and early adopters and secondly, on the link between two key actors, namely private sector companies and public policy-makers. Similar approaches are taken by Weber et al. (2009) and Leitner et al. (2010), who both closely link the process of diffusion to the policy framework.

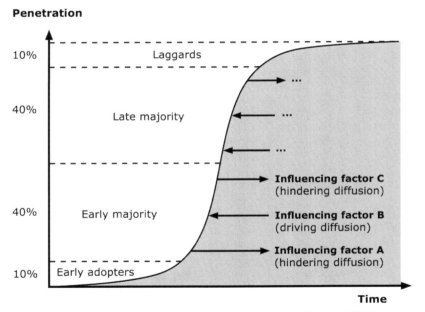

Figure 1.1: Diffusion of innovations, adapted from Rogers (1995)

Following the work of Metcalfe (1988), the literature surrounding diffusion can be divided into two streams. The first attempt to characterise mechanisms and patterns of diffusion and focuses on the rate and amount of adoption in a given population within a time period. The second revolves around the understanding and characterisation of the decision-making structures and processes that influence product adoption, focusing on individual decision-making. Sarkar (1998), Geroski (2000) and MacVaugh and Schiavone (2010) further elaborate on both approaches.

An aggregated analysis of the diffusion of large and small scale energy infrastructure is very useful to gain a general understanding about energy markets and related policy mechanisms. One vital observation is the high dependence of the diffusion curves on changes within the social system (Wong, 2005; Späth and Rohracher, 2010; Dewald and Truffer, 2011). In the context of large scale energy infrastructure, aggregated market data is mostly used to analyse and compare policy mechanisms and to discover which one would best support market deployment in existing markets (Jacobsson and Bergek, 2004; Buen, 2006; Ringel, 2006; Kristinsson and Rao, 2008; Mostafaeipour, 2010). However, it is important to note that the support mechanism by itself is a rather narrow conceptualisation and that one cannot fully explain deployment, even in well-researched industrialised countries. In the context of small scale renewable energy, most studies focus on one or several countries or technologies (Hammond et al., 2007; IEA, 2010; Chaurey and Kandpal, 2010). These studies, based on estimation methods and academic literature, produce high-level recommendations for national and international policy-makers and companies.

Thus, the micro-perspective of individual decision-makers potentially results in a deeper understanding of diffusion patterns. This conclusion is in line with Metcalfe (1988) who clarifies that this approach allows the exploration of why and when adoption might occur. Here, models and approaches available include a wider array of factors that arguably influence the decision to adopt. In the field of large scale energy infrastructure, this approach was first proposed by Wiser and Pickle (1998) who concludes that many policies are not effective in terms of market growth, as they do not meet the needs of investors. Later, both qualitative and quantitative approaches identified the effect of policies on perceived investment risk and related financing costs that affect market deployment (Butler and Neuhoff, 2008; Gross et al., 2009; Sovacool, 2010; Lüthi and Prässler, 2011). In the context of small scale energy infrastructure, researchers have conducted, on the one hand, surveys and interviews with end-users (McEachern and Hanson, 2008; Ndzibah, 2010; Abdullah and Mariel, 2010; Sovacool et al., 2011; Rebane and Braham, 2011) and on the other, qualitative case studies of individual companies (Lemaire, 2009; Anderson et al., 2010; Mukherji and Jose, 2010; Prahalad, 2012). All these methods expose a rich variety of contextual factors such as knowledge, awareness, finance and social aspects.

In summary, aggregated analysis seems a suitable approach in existing markets where historical data on market development is accessible. However, in nascent markets of emerging and developing countries, this kind of historical data does not yet exist. In order to draw valid conclusions in the latter context, it is necessary to identify and evaluate the adoption decision at the micro-level.

1.3 Research context and research question

At an abstract level, the research context of the deployment of renewable energy technologies is determined by several factors (see Figure 1.2). On the one hand, natural resource limitations, such as inadequate solar radiation or reduced wind speed, would hamper the technical feasibility of either technology in a given region. On the other hand, social acceptance for these new technologies can create different effects on their deployment. Within these broad limitations, three elements, namely policy, finance and technology, are interconnected. Smart policies lead to an increased application of the technology. As a result, the technology is improved, for example due to learning curve and scale effects, which increases its attraction for the investor. Social acceptance as well as policy and finance are influenced by global trends such as climate change and increasing fossil fuel price volatitily. This thesis follows the assumption that in a broader sense global challenges and social acceptance are in favor of renewable energy technologies. It is also assumed that natural resources such as wind and solar radiation are available to a sufficient degree and therefore do not require further consideration.

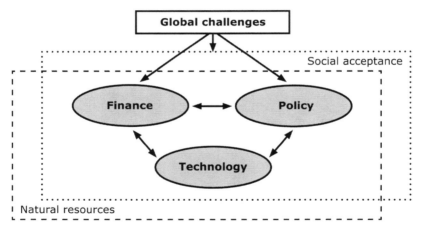

Figure 1.2: Conceptual framework for renewable energy technologies

While energy infrastructure can be applied in different regions of the world (e.g. developing, emerging and industrialised countries), it can be designed in different ways (off-grid and grid-connected). In the context of renewable energy technologies, decentralised designs, while willing to compromise on the quality of the natural resource, play a vital role in minimising transport losses and dependency on other regions or countries[1]. In contrast, centralised designs apply renewable energy technologies in ar-

---

1   e.g. photovoltaic energy in Germany

eas where natural resources fit perfectly with the needs of the technology, accepting transport losses and dependency even across national borders[2]. Some combinations of infrastructure design and regional context match well while others do not seem feasible. For example, off-grid applications make sense only in very few applications within industrialised countries, such as remote buildings in mountainous regions. However, the same design makes a lot more sense within developing and emerging countries, as many potential consumers are not yet connected to the grid.

In line with economic growth, the energy demand in emerging and developing countries is rising quickly, increasing the resultant demand for new energy infrastructure. The increasing need for energy supply is an opportunity for these countries to make a partial or complete jump to renewable energy technologies, thereby bypassing some of the "dirty stages of development experienced by industrialised countries" (Watson and Sauter, 2011). While Soete (1985) discusses the issue of leapfrogging at a more abstract level, Chircu and Mahajan (2009) analyse the market penetration of mobile phones in BRIC countries compared to industrialised countries and found that the existing gap can be narrower, absent, or even reversed. In the case of power generation, leapfrogging directly to renewable energy technologies would allow emerging and developing countries to immediately capture the benefit from reliable and predictable energy prices as well as other advantages such as independence from international fossil fuel prices and local job creation. Therefore, this thesis focuses on emerging and developing countries as an area of application for renewable energy technologies.

Investigating the effective and efficient application of renewable energy technologies in emerging and developing countries uncovers several perspectives, and each raises a multitude of questions. For example, one could analyse either the economic point of view by considering the optimum welfare distribution of different power generation technologies within the energy system of emerging countries, or examine the policy perspective and motivational factors for developing different framework conditions for (renewable) energy infrastructure investments. While these perspectives are important, this thesis limits its focus to early adopters of the technology, therefore highlighting the viewpoint of the private sector with regard to framework conditions and other related aspects (see also Chapter 1.1). As one of the key stages of technology diffusion relates to early adopters or risk-taking entrepreneurs (Rogers, 1995; Patzelt and Shepherd, 2009), the following research questions seem relevant in revealing key elements of the diffusion process:

RQ 1: Which influencing factors hinder early adopters in diffusing renewable energy technology within emerging and developing countries?

---

2   e.g. large hydro dams, off-shore wind farms

As the context is, by definition, different for grid-connected and off-grid renewable energy technologies (see also Figure 1.3), the second and third research questions focus on issues within each infrastructural context individually. In the case of grid-connected power plants, there is a strong interdependency of private companies with regard to regulation and the grid-operator. Therefore, the analysis concentrates on international private sector companies and investors and their perception of national regulation and other influencing factors:

RQ 2: How are different influencing factors evaluated by early adopters of grid-connected renewable energy technology in a highly regulated emerging market?

In the context of off-grid power plants, the link between technology and the end-consumer is critical to success, as the spatial distance between the two is virtually non-existent. This includes not only the technology but potentially refers to other services such as maintenance and finance. Policy-makers can facilitate the deployment of renewable energy, although past efforts show mixed results (Acker and Kammen, 1996; Sebitosi and Pillay, 2005; Wamukonya, 2007). Therefore, the focus of investigation is not on framework conditions but on the link between company and consumer.

RQ 3: How are different influencing factors evaluated by early adopters of off-grid renewable energy technology, considering their interaction with low-income consumers?

By focussing on specific key aspects RQ 2 and RQ 3 complement and refine the findings from RQ 1. An overview of the value chain and the specific focus of the three research questions is provided in Figure 1.3.

1.4 Case study method

This thesis aims to shed light on the application of established technologies in a new context. An explorative research design is chosen in order to reveal and analyse the perception of early adopters in this new regional context. Using the case study approach allows to study real-world conditions, perceptions of decision-makers and different contextual conditions (Yin, 2011). Thereby the case studies typically combines multiple sources of empirical evidence (Eisenhardt, 1989; Moran-Ellis et al., 2006).

Cases are selected based on to the purpose and empirical context of the study (Eisenhardt, 1989; Seawright and Gerring, 2008). Therefore, one technology for each empirical context (grid-connected and off-grid) is chosen. In order to explore the raised

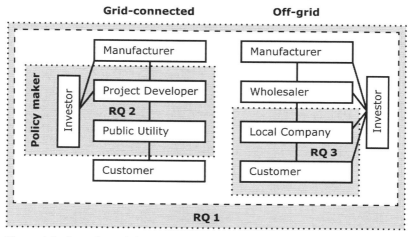

Figure 1.3: Overview of the value chain for grid-connected and off-grid power plants and the focus the research questions

questions firstly, the application of the technology in many markets must be possible and secondly, it is also important to abstract from technological risks as this could influence the results considerably. Therefore, technologies that are mature (with low and well-known technical risks) and that have a high market potential (in terms of available natural resources and potential sites for application) have been selected[3]. Analysing the most typical technology cases allows, to a certain degree, the formation of conclusions that are applicable to the larger group of grid-connected and off-grid power generation technologies (Flyvbjerg, 2006). Thereby, this thesis builds upon three different sources of empirical data, presented in Table 1.1 and enumerated below:

1. Qualitative interviews with private sector decision-makers - typically top managers or company owners - and other experts of the sector were conducted in order to gain a deeper understanding beyond concepts already studied and discussed in the academic literature (see also Table 5.4 and Table 5.5);

2. A quantitative survey with private sector decision-makers was conducted. For both technology case studies, a sample of decision-makers, with proven professional track records and personal experience with the technology in the context of emerging and developing countries, was selected (Chapter 3.3 and Chapter 4.4). In addition, survey methods selected were those suited to small sample sizes and low influence of scale use bias; and

---

3   for more details see Chapter 2.2.1

3. For each technology case study, a workshop with decision-makers and other experts was organised in Germany. The aim of the workshop was to discuss and reflect the survey and interview results in order to refine and deepen the understanding of each technology within its context of application.

Combining these different empirical data sources allows the enhancement of the validity of the explorative study design. While the first research question is answered based on qualitative interviews as well as both workshops, one for each technology focus, the second and third questions each combine qualitative data and a quantitative survey (see also Table 1.1).

The rest of the thesis is structured as follows: The two technology case studies are used in order to firstly understand the challenges and barriers for technology deployment in general (see also Chapter 2 (RQ 1)) and to secondly investigate in more detail, the questions that are key for each technology (see also Chapter 3 (RQ 2) and 4 (RQ 3)). Finally, conclusions are drawn for researchers, policy-makers and managers (see Chapter 5).

Table 1.1: Overview of the empirical data of this thesis

	Case 1: Wind Farm	Case 2: Solar Home System
1. Interviews		
Private Sector	4	4
Other experts	4	1
2. Survey		
Method	Maximum Difference Scaling	Adaptive Choice Based Conjoint
Population	188 decision-makers	93 decision-makers
Response rate	41 responses, 26 %	31 responses, 33 %
3. Workshop		
Participants	13	8
4. Interviews		
Private Sector	4	5
Other experts	-	3

2 Renewable energy diffusion in emerging markets – The micro perspective on barriers for wind farms and solar home systems

Christian Friebe, Paschen von Flotow

Abstract
Currently, the energy sector in many countries is facing a major transition to low-carbon technologies. The context of emerging countries is especially relevant as their domestic energy demand is increasing at high rates, which contradicts with limited government budgets to support renewable energy technologies. To address this gap, international measures are taken to leverage private capital for renewable-energy infrastructure investments in emerging markets. The aim of this paper is to reveal the micro perspective of private sector decision-makers (i.e., early adopters) on barriers to technology diffusion and potential measures by analysing two case studies, namely wind power (grid-connected) and solar home systems (off-grid). The key findings are, firstly, the importance of capacity building and increased interaction and discussion among public and private sector decision-makers, secondly, the need for well-targeted guarantees in the case of grid-connected power plants and thirdly, the need to design and implement suitable measures for refinancing in the case of off-grid power plants.

2.1 Introduction

While there is widespread agreement among researchers and practitioners that favourable policies drive renewable energy and clean technology deployment, what exactly
"favourable" means is still debated (Skidmore, 1997; Bazilian et al., 2012). Even in
the well-studied context of industrialised countries, policy-makers often set regulations
without approved evidence of the potential effects on private companies and investors
(Wiser and Pickle, 1998; del Rio González, 2005; Schleicher-Trappeser, 2012). In line
with Guerin (2009), this paper focuses on the context of emerging markets in emerging
and developing countries and sheds light on the view of private sector decision-makers
regarding perceived barriers to diffusion. We decided to abstract away from technological constraints that are studied by other researchers such as Schilling and Esmundo
(2009) and Hendry and Harborne (2011) and instead to focus on the remaining influencing factors for technology diffusion. The research question of this paper is the
following: What barriers do private sector companies (i.e., early adopters) in the field
of renewable energy technologies face in emerging and developing countries? In addition, the paper also discusses several measures to address the identified barriers to
diffusion.

According to Rogers (1995), diffusion is defined as "the process by which an innovation is communicated through certain channels over time among members of a social system". The innovation can take several forms such as a new process, an idea
or a new product. In this paper, the innovation is technically mature power generation infrastructure in a country with little or no legacy of applying this technology.
The institutional framework interacts with the adopters of a technology (Porter, 1990;
Weber et al., 2009; Leitner et al., 2010). In the case of large power stations, the potential adopters are the national utility or industrial energy consumers. In the case
of small distributed power stations, the adopter could be a municipality or a single
household.

Broad technology trends or mature markets that already existed for some time can
be studied by analysing aggregated market data or by conducting longitudinal studies
(e.g., Schilling and Esmundo (2009); Holt (2011)). However, due to a lack of data,
this is not possible in nascent markets that do not have historically deployed markets.
In the latter case, analysing the micro perspective of individual decision-makers, e.g.,
early adopters, is suitable for gaining a deep understanding of barriers to technology
diffusion. This conclusion is in line with Metcalfe (1988), who clarified that the micro
perspective on diffusion allows researchers to explore why and when adoption might
occur.

In the field of large-scale energy infrastructure, the micro perspective was first proposed by Wiser and Pickle (1998), who found that many policies do not meet the needs

of investors and consequently lead to non-existent or unsustainable market growth. Later, qualitative and quantitative approaches evaluated the effect of policies and different market designs on perceived investment risk and related financing costs that affect technology deployment in industrialised countries (Markhard and Truffer, 2006; Butler and Neuhoff, 2008; Gross et al., 2009; Lüthi and Prässler, 2011). However, only a few studies focus on nascent markets and consequently apply a micro perspective. Komendantova et al. (2009) analysed the application of concentrated solar power (CSP) in North Africa and found that regulatory risks are perceived to be both most consequential and most likely to occur. Political stability and force majeure seem to be somewhat less important. Sovacool (2010) evaluated the perspectives of decision-makers in Southeast Asia to reveal their perspectives on barriers for market deployment. He found that, among others factors, subsidies for conventional energy sources and non-technical barriers are hindering the ability of companies to develop and invest in renewable energy infrastructure.

In the context of small-scale energy infrastructure, researchers so far have focused on different levels of analysis. Firstly, there are studies that focus on the end-user. These papers conclude that affordability, financing and accessibility are keys to success (Ndzibah, 2010). In addition, the social structure of the village has a key influence on adoption decisions (McEachern and Hanson, 2008) and consumers might also prefer to invest their money in products other than a reliable energy system (Sovacool et al., 2011). Secondly, studies that focus on individual companies either analyse the broader context of companies in low-income markets (Schrader et al., 2011) or focus especially on energy services, public-private partnerships and related business models (Lemaire, 2009; Mukherji and Jose, 2010). Finally, some studies analyse whole countries and market trends. On the one hand, these studies can be more conceptual such as Prahalad (2012), who develops a framework focusing on awareness, access, affordability and availability. On the other hand, some researchers focus more on empirical data derived from either simulation tools (Hammond et al., 2007) or public support programs (Ketlogetswe and Mothudi, 2009; IEA, 2010; Chaurey and Kandpal, 2010).

To summarise, this study aims to integrate the micro perspective of decision-makers of individual companies regarding barriers and measures for market deployment in a broader sense. Therefore, this study contributes to the debate of efficient and effective interaction among public and private sector decision-makers. The remainder of the paper is structured as follows: The introduction is followed by a section on the methods and data (2.2). Then, two technology case studies are presented (2.3 and 2.4) and discussed (2.5). The paper concludes with implications for policy-makers (2.6).

2.2 Methods and data

2.2.1 Selection of technology case studies

In general, cases are not randomly selected but are selected based on the purpose of the study and the empirical context (Eisenhardt, 1989). Within this frame, typical, diverse, extreme, deviant, influential, most similar and most different cases can be selected (Seawright and Gerring, 2008). The aim of this paper is to explore barriers to diffusion of proven technologies that can be directly influenced by stakeholders at different levels. In addition, we limit the study to technologies that can be applied in most regions of emerging and developing countries. This approach also allows for discussions of the similarities and differences between typical grid-connected and off-grid power generation technologies in a broader sense (Flyvbjerg, 2006). As discussed in the above mentioned literature only a limited number of case studies exist in the nexus of policy and finance.

In the case of grid-connected renewable energy technologies biomass, conventional geothermal, hydro, photovoltaic and on-shore wind moved beyond the inception phase and are now in either the take-off or consolidation phase (IEA, 2011). We excluded hydro power from the list as this technology requires geological conditions that are only available at very few sites compared to all other technologies. Within the set of the remaining technologies, wind power is the most cost competitive in terms of power generation costs (IEA, 2011; REN21, 2012) and is therefore selected as the first technology case study (Table 2.1).

Table 2.1: Comparison of the selected technologies in the context of emerging and developing countries (REN21, 2012)

Context	Technology	Typical Size	Typical Costs
Off-Grid	Solar Home System	20 - 100 W/system	40 – 60 USct/kWh
Grid-connected	Wind Farm	1.5 - 3.5 MW/turbine	5.2 – 16.5 USct/kWh

In the case of off-grid renewable energy technologies, small wind turbines, photovoltaic and small hydro are in principle able to cover basic energy needs of one household. Depending on the context and the available natural resources, small wind turbines and small hydro can be feasible. However, companies and public authorities have so far focused mostly on photovoltaic systems, although the technology is comparably expensive (REN21, 2012). The main reason for this preference is not the cost of power generation but rather the availability of solar irradiation in most emerging and developing countries. Therefore, photovoltaic (solar home system) is selected as the second technology case study (Table 2.1).

2.2.2 Data sources

To identify key barriers for each case, qualitative data are analysed. These data include interviews with experts and private sector decision-makers, as well as one workshop for each case study to share and discuss the results of the interviews (see Table 2.2). The interview data consist of notes taken during and directly after the interview. In line with Bewley (2002) and based on the initial interviews, we found that this approach allows the interview partners to reveal more insights and speak more openly compared to a tape-recorded interviews. Each interview is carefully prepared by analysing publicly available information on the company and the interview partner. This procedure allows us to adapt our key questions (see Table 5.3) to the specific circumstances and regional focus of the company and interview partners. Based on the results of the interviews, two workshops have been conducted that allowed for a direct interaction of the decision-makers. This second source of qualitative data is used to strengthen and refine preliminary findings from the interviews.

Table 2.2: Overview of the number of interview partners and workshop participants (see also Table 5.4 and Table 5.5)

Category		Case 1: Wind farm	Case 2: Solar Home System
Interviews	Private sector	8	9
	Experts	4	4
Workshop	Private sector	6	4
	Expert	7	4

To identify interview and workshop participants, we followed a purposive sampling strategy, as suggested by Yin (2011) and others. With one exception[4] all participants belong to different organisations or companies and each involved private sector company is currently active in at least one emerging or developing country. All participants from the private sector have personal experience in at least one emerging or developing

---

4   The participants OWG3, WIG8 and WWG5 do belong to the same holding company. However, OWG3 and the other two participants belong to different and completely independent subsidiary companies, while WIG8 and WWG5 provided different perspectives on one subsidiary company (top management and financial expertise). The three persons were contacted individually during the data-gathering process (no snowball sampling).

country and are either owners of the company or belong to the top management. Additional experts are interviewed to add a different perspective. Interviews were conducted until theoretical saturation had been reached, as suggested by Eisenhardt (1989). In a process of several iterations, key barriers for the private sector were identified and several higher order groups of barriers emerged.

In the case of wind farms, mostly project developers from Germany are involved in the study (Table 5.4). This is reasonable from our perspective as firstly, German project developers are already very active in the international context (compared to other countries) and secondly, our focus of analysis is not the company but the perspective of the private sector decision-maker regarding energy markets. Thus, India and China, as two major wind markets, are explicitly not part of this study. We argue that both markets have already moved beyond the initial stages of market development and focus on remaining emerging markets in Eastern Europe, South America and beyond (Table 5.4). In the workshop, the participants are not only project developers but also cover other parts of the value chain (see also Figure 2.1), which allows us to include a broader perspective and, to some degree, to triangulate the results from the interviews.

In the case of solar home systems, the workshop participants are working for wholesalers of systems, while interview partners are from local companies in two different countries, namely India and Tanzania (see also Figure 2.2). We selected these two countries as they cover different cultures (Africa and Asia) and different stages of market development. While India is already today one of the largest markets for off-grid electrification in Asia and beyond, Tanzania is currently in the early stage of market development in Africa (Hammond et al., 2007; Byrne, 2011; Ondraczek, 2012). Thus, we could cover two different parts of the value chain while including personal interviews with decision-makers who have different country perspectives. Again, this allows us to triangulate the results to some degree (see also Table 5.5).

Within the case studies, we refer to the perceptions of the decision-makers by using a simple code that refers to the list of experts in the appendix (Table 2.2 or 5.4 and Table 5.5). The first letter refers to the technology (*O*ff-grid or *W*ind), the second letter refers to the mode of interaction (*I*nterview or *W*orkshop) and the third letter refers to the country of origin of the company (*G*ermany, *T*anzania, *I*ndia). If the participant is an industry expert on the topic, we used a different coding (*E*xpert).

2.3 Deployment of wind farms (case 1)

The value chain for wind farm projects (Figure 2.1) includes manufacturers that build the components such as tower elements, nacelle and blades. Project developers have a competency in analysing market conditions, policy frameworks and negotiating with different actors such as land owners, manufacturers, investors, policy-makers and the national utility. Once all approvals are available, the construction and commissioning of the wind farm can start. To obtain these approvals, project developers very often establish local subsidiaries or joint ventures. National and international private investors evaluate wind farms to be a suitable long-term investment depending on their respective risk appetites. Beyond investments in wind farms, other actors might have additional refinancing needs. As the electricity grid is vital for economic development, policy-makers at the national or international levels have a key influence on the deployment of wind farms by applying different measures. For the case of wind farms, seven barriers emerged from the data. These barriers are often interlinked. In what follows, each barrier is discussed in detail.

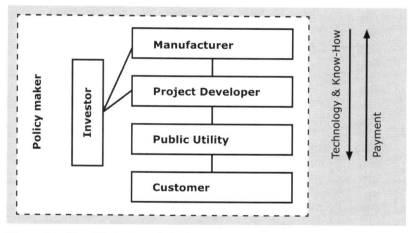

Figure 2.1: Simplified value chain for wind farm projects including key stakeholders

Firstly, the level of uncertainty regarding electricity sold to the utility can become a major barrier (WIG5). This issue can be related both to the missing financial stability of the off-taker, which could mean that additional government guarantees are required to make a project bankable (WIE2, WIG2, WIG3) and to the tariff that is paid for power generated by the wind farm. More specifically, electricity tariffs can be fixed (e.g., a power purchase agreement or a feed-in tariff) or flexible, depending on the demand and supply of electricity (e.g., a quota system). In the first case, private investors know about the electricity tariff, which reduces risk and, consequently, return expectations (WIE4, WIG3, WIG4, WIG7). In the second case, the electricity tariff is a considerable risk for the investor, which often leads to non-bankable projects (WIG5). In fact, a feed-in tariff significantly reduces the risk for investors and is therefore especially preferred by several decision-makers (WIG1, WIG2, WIE3, WWE3), while other mechanisms such as a quota system increase risk and, consequently, return expectations of investors (WIE4). In addition and depending on the national context, public authorities might decide to adjust tariffs for inflation or currency risk, which considerably reduces investment risk (WIE4, WIG2, WIG4, WWG5, WIG6). These types of risk-reducing guarantees can also be offered by international public authorities, e.g., as part of development aid or international climate finance (WIG1, WIE4). In a broader sense, several decision-makers call for a transparent long-term public financing strategy for renewable energy support mechanisms (WIE4, WIG1, WIG3). Interestingly, some policy-makers attempted to kick start their renewable energy markets with very attractive feed-in tariffs. For example, in eastern European countries, many governments announced comparably high tariffs resulting in a "gold rush" by project developers (WWE2). However, as the approval process was not well defined, no project developer obtained the necessary approvals, as policy-makers decided to significantly reduce the tariffs and define stricter criteria for the approval process (WWG1, WWE2, WWE3). This consequently led to high perceived uncertainty by investors (WIG7) and to the second barrier to technology diffusion.

Secondly and beyond the debate of which support mechanism is most effective, the proper implementation of the chosen mechanism seems to be even more important for private sector decision-makers. While some interview partners and workshop participants highlight that the process of getting permits is not defined (WIG2, WIG5, WIG7), others see a major barrier in the implementation of a defined process (WIE3, WIG4, WEG6, WIG7). In this context, legal security, e.g., the option to sue authorities if they do not follow the decided process (WIG1, WIE1, WIG2, WIG3) and the level of corruption at the national, district or municipal level (WIG1, WIG6, WIG7, WIG8), becomes key. Tenders can include a specific risk: Although positive examples exist (e.g., South Africa), tenders for which decisions are made behind closed doors might significantly increase the risk for unfair practices (WWE2, WIG7, WIG9). All named aspects influence the duration of the approval process, which is important for market development and private sector activities (WIE1, WIG4). Capacity building, potentially financed by international donor agencies, can help to address this barrier (WWE5). However, the good intentions of donor agencies can also lead to non-sustainable results, as one project developer highlights, "Who to trust? Unqualified[5] consultants paid by donors or companies that are looking for projects?" (WIG8). Potentially, this lack of capacity can lead to, for example, unrealistic local content requirements (WIG8, WIE4, WWE2), tendering documents that are not suitable (WIG8)

---

5   The interview partner defines "unqualified" in this context as a lack of knowledge and understanding regarding the private sector perspective.

or one-stop shops that do not have sufficient institutional power to truly accelerate the approval process (WWE2, WWE4). One can conclude that addressing these barriers entails defining and implementing a suitable approval process step by step to build up the trust of investors. Therefore, it is a good sign if, firstly, the government publishes information regarding, for example, the national energy market, a national plan for renewable energy, rough estimations of the wind speed at different locations and the approval process itself (WIG1). Secondly, public policy-makers could build up trust among private sector decision-makers by focussing on transparency and predictability if changes to the policy framework are necessary (WIG1, WIE1, WIE2, WIG3, WIE4, WIG7).

Thirdly, interfacing with the grid infrastructure is often a challenge for all involved stakeholders. This can be an organisational issue such as the priority of power dispatch within the grid (WIG8) or a technical issue such as the reliability and the capacity of the grid (WIE3, WIG2, WIG4, WIG5, WWE7). In fact, grid access is, by definition, required for grid-connected power plants and should be a key element of any well-defined and implemented process (WIE1, WIE4, WIG2, WIG7). Grid access and power dispatch are typically regulated by public authorities and are often guaranteed to project developers or offered as part of a power purchase agreement. Therefore, facilitating grid access by the capacity building of policy-makers, regulators and the national utility can be a major lever by which to drive renewable energy diffusion (WWE5, WIG7).

Fourthly, high local content requirements can become a major challenge or barrier to diffusion. Many policy-makers aim to build up a national industry for renewable energy power plants to create local jobs. However, as nascent markets often experience unstable market development at comparably low levels, only a few international manufacturers are able to deliver their wind turbines and related after-sales services (i.e., maintenance) (WWG6). Engaging international manufacturers to locally source components such as tower elements or even build up local facilities is challenging in nascent markets (WWG6). In addition, in many cases, turbines sold in these markets are not the latest technology that is available in more mature markets (WIG5). Therefore, limited access to wind turbine technology can become a barrier to technology diffusion. Policy-makers are often tempted to address this issue by including local content requirements in the national regulation, which creates other challenges. In fact, high required shares of local content in nascent markets often lead to a significant delay in market development – a side effect that is desired by neither policy-makers nor the private sector (WIG8, WIE4, WWE2). As a solution, private sector decision-makers suggest addressing this barrier by, firstly, creating a stable market development and, secondly, implementing local-level content requirements that can be met by the existing local industries (WIG8).

Fifthly and in line with the core business of project developers, it is a challenge to build and develop local networks and partnerships with key stakeholders to, for example, acquire suitable land or to push projects through the approval process (WIG3, WIG4,WWE3, WIG8). Therefore, personal contacts or a local partner company are keys to success (WIG1, WIG2, WIG3, WIG4, WWE5, WIG5, WIG7).

Sixthly, it is often challenging for project developers to secure potential sites with suitable wind resources. Typically these areas are rented from a local farmer, which sometimes requires solving difficult contractual situations or includes the challenge of identifying the formal owner(s) of the area(s) (WIE4, WIG5, WIG7). However, once the contracts are signed and the project developer moves on to wind measurements and applications for the approvals, a considerable competitive advantage over competitors is realised (WIG6).

Finally, the availability of capital is a challenge and can turn into a major barrier, as the risk that results from the previous barriers is often perceived to be too high by many investors and banks. The balance of risk and return is a required key competency of project developers, investors and risks (WIE3, WWG3). However, the ability to handle especially high perceived risks varies considerably depending on the investor and the project developer. One project developer who just closed down all activities outside Europe noted, "regarding risk and return, emerging countries are competing with low-risk, low-return, quick-approval-process projects in Germany and the EU" (WIG1). Public international investors such as donors are key for the first projects and beyond – especially after the financial crisis (WIE4, WIG7). However, the risk appetite of public investors is also limited as both a project developer and an expert for public finance highlighted: "donor funding is available – in fact, donors compete for good projects, but the number of good projects is limited" (WIE1) and "oftentimes development banks are not willing to take more risk than commercial banks" (WIG5). In this case, local private or public banks might be much better able to deal with political risks (WIG1, WIG3, WIG4, WWG1, WWG4). One project developer clarified, "during the first five years, you always have to secure funding locally" (WIG6). However, local investors and credit agencies often lack the capacity to do the due diligence for wind farm investments. This circumstance can also be addressed by partnerships with other private investors such as utilities that use balance sheet finance (WIG8) or large local cooperation that aim to diversify their portfolio with equity investments in wind farms (WIG5, WIG8). In summary, one can conclude that the higher the risk for the private sector, the more difficult it is to reach financial closure. It seems as if low risk and low return are currently preferred by most investors, with the exception of local private equity investors. Therefore, well-targeted country-specific guarantees regarding, for example, tariffs, inflation or currency that are provided by national or international public authorities can significantly increase the deployment of wind farms (WIE4, WIG6).

2.4 Deployment of solar home systems (case 2)

The value chain for rural off-grid electrification (Figure 2.2) includes manufacturers that build the components such as the inverter, the photovoltaic module and the battery, as well as wholesalers that have a strong competency in combining different components to create a functioning power supply system. While manufacturers and wholesalers are typically active in the international market, the local company directly interacts with the consumer who buys and uses the off-grid energy system in the village or household. Depending on refinancing needs, all actors interact with public or private investors that facilitate the flow of technology and know-how from manufacturer to consumer. Policy-makers at the national or international level can influence the deployment of off-grid power supply with different measures. For the case of solar home systems, five potential barriers emerged from the data. In the following, each barrier and its potential link to other barriers is discussed in detail.

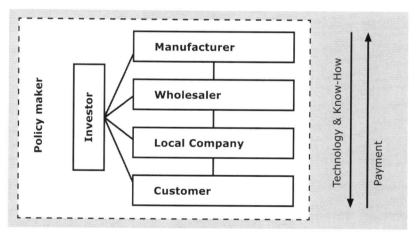

Figure 2.2: Simplified value chain for off-grid projects including key stakeholders

Firstly, the commitment and capacity of policy-makers emerged as a major barrier to technology diffusion. In fact, private sector decision-makers feel a lack of commitment from public authorities for rural electrification. Interview partners claim that policy support often focuses on grid extension and considers off-grid solutions as not "serious" (OIE1, OIT1, OII4). Raising limited public funds for rural electrification projects in this context is challenging for public authorities who are willing to support the sector (OIE3). Based on this perception, workshop participants call for a fair and transparent national policy framework for off-grid electrification (OWG2, OWE4, OWG3). Even if sufficient commitment among public sector decision-makers is available, there seems to be a lack of policy-making capacity for designing and implementing suitable framework conditions at the national, district and municipal levels (OIE1, OIT3, OIE2, OIE4, OII3). The lack of competencies at the policy level can lead to tender specifications for grants that disturb the market in a non-sustainable way (OIE2, OIT2). For example, some government tenders do not include maintenance service, which leads to a high number of broken systems and a bad reputation for solar home systems in general (OIE4, OII3, OIT2). In addition, some companies lack the skill to successfully bid for tenders (OIE1, OIT5), while other companies are tempted to offer low quality products to win the tender (OIE2, OIT2). One interview partner concludes, "don't waste your time on government tenders" (OIT2). Policy-makers sometimes decide to allocate grants for the customers of solar home systems or the companies that provide these systems. However, the interview partners and workshop participants stated that this measure is in most cases disturbing the market and competition. Therefore, they are suggested to focus less on grants and more on investments (OII2, OII4, OIT1, OWG4). If the government would like to apply grants in the rural energy sector, this measure should not be available for consumers who buy solar home systems (OWE2, OWE1) but should be well targeted to finance the following:

- feasibility studies (OIE3, OWE2, OWE3)

- demonstration plants on public buildings (OWG1)

- support maintenance activities (OWE1)

- due diligence for private investments (OIT1, OWE2, OWE1)

It is suggested that a suitable policy framework for off-grid electrification could possibly support the sector by allowing for an import tax and VAT exemption for all renewable energy technology imports (OIE1, OIT1, OIT2), government guarantees for facilitating the import of technology (OWG2) and equally distributing or equally cutting subsidies for grid-connected electricity and off-grid electricity (OWG2, OWE4). One suggestion by an interview partner is to set up a voucher system that allows low-income households to use them for either buying fossil fuels or paying back their solar home system (OII4).

Secondly, interview partners highlight the lack of skilled employees – even if employees obtained a formal education (OIE1, OIE2, OIE3). Although the unemployment rate is often high, it is very difficult for companies to identify and hire skilled employees (OIE1, OIE3, OIT2, OIT4). To address the lack of skilled employees, cooperation between industry, research and policy to develop a suitable official certificate for proper training in, for example, solar home system installation could tackle both industry needs and opportunities for employment (OIE2).

Thirdly, regarding the technology supplier (manufacturer or wholesaler), a key challenge and potential barrier for market deployment is competition from companies that offer low quality products (OII4, OWG3). Lower quality allows one to offer lower prices, which is not only relevant for the interaction with consumers but also for winning public tenders. The effect is reinforced by some companies labelling their products in a misleading way (OIE2, OIE1, OIT4). Therefore, the interview partners and workshop participants suggested developing a quality label or a reliable minimum standard for the technology (OIT4, OWE2). This could be implemented for government tenders. So far, government tenders seem to focus more on low prices than on good quality. As one interview partner underlined, "the one who wins the tender goes with low quality products" (OIT3).

Fourthly and in line with the quality issue, the low reliability of the solar home system due to the lack of adequate maintenance can be a barrier. Beyond product quality, reliability requires interaction with customers regarding education (before selling a solar home system) and maintenance services (after selling a solar home system). Many companies that have a long-term interest in the rural energy market feel responsible for helping customers to understand the benefits and limits of a solar home system before the customer can make an informed buying decision (OIE1, OIE3, OII3, OIT2). This arrangement could also include a one-week testing phase at the customers home (OII1). After the customer has bought the solar home system, a minimum of maintenance service is required. Delivering this service on a commercial basis is challenging for many entrepreneurs, especially due to the distance between the company and the customer (OWE1, OWE2, OW4, OII1, OII4). To conclude, maintenance services are necessary for succeeding in the long-term, although currently, it is challenging for companies to provide these services due to the high cost (OIT1, OWE1).

Finally, most customers that intend to buy a solar home system face the challenge of affordability or limited purchasing power. Therefore, financing services are necessary but often challenging to implement in the local company (OIE1, OIT1, OIT4, OIE2, OII2). Microfinance organisations can potentially address this barrier, as they have already established a good relationship (i.e., financial track record) with potential customers for solar home systems (OII3, OWE1). However, microfinance organisations and other local banks often find it difficult to adapt their credit products to the financing of solar home systems (OIE2, OIE4, OWE1). One can conclude that, depending on the context, the results of microfinance and renewable energy are mixed. Therefore, many local companies try to offer financial services to their customers. However, local companies often have very limited access to capital sources. In fact, many interview partners and workshop participants raised the issue that local companies have a relevant refinancing need to develop their business (OIT1, OIT2, OIT3, OIT5, OWE1, OWG1, OWE2, OWG3, OWG2). Offering financial services such as a 2-year repayment period seem to be key to success in the rural energy context by both local companies and international wholesalers. The latter delivers mostly on a prepayment basis, as wholesalers are reluctant to take the risk of non-payment from their local counterpart (OWG1). Developing the capacity of manufacturers and wholesalers to possibly work jointly with export credit agencies and other partners on new approaches to support local companies with their working capital needs, e.g., with adjusted payment conditions, could be a first step to addressing this barrier (OIT5). Regarding investors that could potentially provide capital, the current needs of the local companies can be only partly met by current funding sources:

- Public investors such as donor agencies look for investment opportunities beyond 10 million USD due to high transaction costs (OIT1, OIT2) or support pilot projects through grants (OWG2, OWE3).

- Private banks lack the capacity to understand and support the rural energy business (OII4, OII1). Due to both high perceived risks regarding country, technology and company and also high costs for due diligence, private banks find it difficult to assess the related risks and provide debt for these companies (OIT1, OWE1).

- International social investors are in principle able to provide debt and equity up to 1 million USD (OIT1, OIT2). However, many of them find it difficult to conduct a due diligence assessment due to a lack of expertise and the high transaction costs.

To go beyond pilot projects and focus on scaling up, the interview partners and workshop participants suggest developing a revolving private or public-private partnership fund that offers (low interest) loans to local companies (OIE1, OIE3, OIE4, OII4, OIT1, OII2, OWG4). Therefore, managing the transition from a publicly funded pilot project to public private partnership funds to the privately funded scale-up of business by local banks is a key to the long-term success (OIE3, OIE4, OWE3). This transition obviously requires stakeholder involvement and training at various levels. Interestingly, one workshop participant raised the issue that, for him, it is currently easier (lower transaction costs) to obtain a grant from a social investor than to raise private and public investments for his projects in Africa (OWG2).

2.5 Discussion

Both case studies reveal specific and similar challenges and potential barriers to technology diffusion. Understanding these similarities and differences can help to derive a more holistic understanding of the diffusion and the rationale of private sector decision-makers (see also Table 2.3). Thus, potential measures concerning public and private sector organisations are discussed. One has to note that all barriers are "bottom-up" and assigned to broader categories, namely support mechanism, infrastructure, business specific and finance. However, one has to remember that many revealed barriers are closely interlinked and cannot be analysed and understood separately.

Table 2.3: Comparative overview regarding potential barriers that are identified for each case study

Category	Case 1: Wind Farm	Case 2: Solar Home System
Support mechanism	Uncertainty regarding electricity tariff	-
	Lack of capacity for implementation	Lack of capacity and commitment of policy-makers
Infrastructure	Unclear interface with grid infrastructure	-
	High local content requirements	Lack of skilled employees
Business specific	Lack of local expertise	Low quality products distort market
	Challenge to secure potential sites	Unprofitable but required maintenance services for customers
Finance	Lack of capital supply for financing projects	Lack of capital supply for providing financing services for customers

Regarding support mechanisms, private sector decision-makers in the field of solar home systems expressed that, oftentimes, there is a lack of commitment by policy-makers. Compared to large grid-connected power plants or grid extensions, the business of solar home systems seems to be somewhat less interesting to policy-makers. Additionally, both cases reveal that further developing policy-making capacity for the design of suitable support schemes and that the implementation of these seems to be a major lever for technology diffusion. So far, the diffusion of solar home systems often faces market distortions, while wind farm construction is often delayed due to, for example, difficulties in obtaining grid access or meeting the local content requirements. Thus, this research complements the work of Sovacool (2010) on support mechanisms and incentives for renewable energy in Southeast Asia. To conclude, this barrier is at the centre of many national efforts and development aid projects. At the national level and in line with Komendantova et al. (2009), this challenge calls for capacity-building measures for policy-makers and local companies, along with stable market development. At the international level and in line with IEA (2010), the identified institutional barrier could also be interpreted as a call for more discussion and inter-action among private sector companies, national policy-makers, donors, development consultants and researchers. The work of Weber et al. (2009) on the diffusion of stock exchanges between 1980 and 2005 suggests that a process that is not donor driven will show better results in terms of vibrant local industry activities.

Regarding the infrastructure, the qualitative data reveal that wind farms often face the challenge of grid access and power dispatch, which is obviously not a barrier for so-lar home systems. If policy-makers aim to build up the local industry, well-adjusted local content requirements for wind farms and high quality training measures for tech-nicians in the case of solar home systems are key for market deployment. Lewis and Wiser (2007) and IRENA (2012) further elaborated on this issue. Interestingly, no lo-cal entrepreneur in the field of solar home systems stated that accessing customers is a challenge. One conclusion would be that, thus far, solar home systems have not yet sold in "deep" rural areas, as discussed by Anderson et al. (2010).

Regarding the business specific barriers for international project developers of wind farms, it emerged that they find it difficult to identify local partners that can, for ex-ample, secure potential construction sites and help to guide projects through the ap-proval process. From the perspective of policy-makers, one can conclude that the need for local partners and expertise creates local employment and fosters the transfer of knowledge. Regarding solar home systems, several companies highlighted the chal-lenge of competing with low-quality products, especially in rural areas. To develop the market, the latter barrier could be addressed both by private sector companies inform-ing potential customers about quality of products (e.g., by showing them low-quality products) and public authorities adopting, for example, import regulations (Jacobson and Kammen, 2007). In line with these barriers and Lindner (2011), the provision of maintenance services seems to be key to success – although, thus far, it seems to be difficult to provide this service on a commercially attractive basis.

For different reasons, private sector decision-makers perceive the access to capital to both wind farm projects and local companies that provide the solar home systems, maintenance and financial services as key barriers to market deployment. In fact, many customers are only able to buy a solar home system if financing services such as re-payment over a period of, for example, two years is provided. In line with Lemaire (2009), providing solar home systems is difficult due to a lack of capital supply for local companies. Compared to wind farm investments, the refinancing need of many local companies supplying solar home systems is quite low. Related to the need for small investments, the transaction costs for local, international, public and private in-vestors are very high in emerging and developing countries. More specifically, the solar home systems sector seems to lack suitable mechanisms by which to allocate capital, while the wind energy sector struggles more with balancing risk and return for investors. This barrier can be addressed, firstly, by policy-makers developing suitable support mechanisms that provide capital rather than grants for companies in the solar home systems sector. Secondly, policy-makers can explicitly reduce the risk for investments in wind farm projects by implementing well-targeted country-specific measures (e.g., guarantees) or with public private partnerships for project finance that address high-risk investments (Wiser and Pickle, 1998; Lüthi and Prässler, 2011; Kleimeier and Versteeg, 2010).

2.6 Conclusion

This paper explores the micro perspective on challenges and potential barriers to technology diffusion. Based on interviews and workshops with private sector decision-makers, renewable energy technologies for both grid-connected and off-grid applications are explored. This paper contributes to an ongoing debate on how to effectively and efficiently support renewable energy deployment in emerging and developing countries. The aim of this paper is to help policy-makers better understand the rationale and priorities of the private sector. Obviously, public sector decision-makers must allocate limited public resources not only based on the needs of the private sector. In fact, national priorities, development aid agendas, economic policy, environmental policy, social policy (e.g., hospitals and education) and economic needs (e.g., reliable electricity supply) all compete for limited public funding (WWE4, WIE1). Balancing these various interests often leads to framework conditions that burden private sector decision-makers in the field of renewable energy. However, the voice of the private sector is helpful to design effective and efficient regulations. Thus, the results of this study are especially relevant to national policy-makers and donor agencies that aim to increase the share of renewable energy in a specific country or region.

For the case of grid-connected renewable energy technologies, the example of wind farms revealed three different levers for public policy. Firstly, it is key to build trust among investors through predictability and transparency in all government activities. Secondly, it is key to have well-targeted instruments such as public guarantees for private investments. One project developer said, "Please do not use the watering can principle but well-targeted guarantees for each country" (WIG5). Thirdly, it is key to do capacity building at the institutional and technical levels to allow for grid integration and streamlined approval processes.

For the case of off-grid renewable energy technologies, the example of solar home systems provides key recommendations to policy-makers. Firstly, focus financial support to companies in this sector more on investments and less on grants. Secondly, choose a support mechanism that does not distort private sector activities and competition in the nascent but evolving markets. Thirdly and in line with the recommendations for the case of grid-connected renewable energy technologies, capacity-building efforts with all involved public and private stakeholders within a country are important.

This paper has several limitations that can lead to fruitful further research. Firstly, this paper covers only the private sector perspective and therefore does not consider all of the factors that policy-makers must take into account while designing and implementing regulations for private investments in renewable energy. Secondly, we could only conduct interviews in two regions for the case of solar home systems and cover only international project developers of wind farms that are based in Germany. This seems to be a serious limitation to the results of this paper. However, even base on these limited empirical base, we can higlight a diversity fo critical barriers. Further research on concrete technical application and their diffusion can help to refine our understanding of these barriers influencing business and investment gaps and thereby technology diffusion.

Beyond these limitations, this paper also presents a small but relevant contribution for international climate change negotiations regarding, for example, the current debate on new facilities such as the the Green Climate Fund. Once established, this fund aims to leverage private capital for and in emerging and developing countries for renewable energy investments. In line with other researchers (Mathews et al., 2010; Bird et al., 2011; Sierra, 2011), this paper aims to provide ideas for potential fund policies. By revealing the perspective of companies regarding renewable energy deployment in emerging and developing countries, this paper shows the needs and challenges for commercially driven early adopters of renewable energy. Several measures for national and international public decision-makers are suggested to effectively and efficiently facilitate market deployment for both grid-connected and off-grid power supply.

3 Exploring technology diffusion in emerging markets – the role of public policy for wind energy

Christian Friebe, Paschen von Flotow, Florian A. Täube[6]

Abstract

This study challenges the implicit assumption of homogeneity in national institutional contexts made in past studies of (renewable) energy policy. We propose that institutional differences matter by focusing on several technology-specific and generic policy factors that can foster technology diffusion through private sector activity. More specifically, we explore perceptions of early adopters in emerging economy contexts using wind park project developers as an example. By applying a parsimonious method for our questionnaire as well as qualitative data we make several contributions: Methodologically, we introduce Maximum Difference Scaling to the energy policy domain. Empirically, we identify several public influences on private investment, and assess their relative importance. This leads to new insights challenging findings from industrialized economies; we identified additional institutional barriers to diffusion, hence, the requirement of a combination of technology-specific and generic policy measures.

---

6   The opportunity to present and discuss earlier versions of the paper at the "EWEA Conference 2011" in Brussels, the "Druid Society Conference 2011" in Copenhagen and the "WWEC 2011" in Cairo helped tremendously in further refining our argument.

For this book the most current version of the paper is included. The original version of the paper that was submited to EBS in August 2012 is available uppon request.

3.1 Introduction

Despite being established technologically, wind power is still relatively uncommon in many emerging markets. As in other infrastructure industries, there is a strong role for policy inducing innovation (Nemet, 2009; Huberty and Zysman, 2010) and to impact its success at the firm level (Lee, 2009). However, it is much less understood how policy makers can influence the step in between, in other words which public measures trigger adoption and diffusion of innovative technologies. Moreover, diffusion of established technologies is to some extent taken-for-granted, yet there are several barriers to adoption, particularly in emerging markets (Popp, 2010; Kemp and Oltra, 2011).[7] In order to better understand this kind of diffusion one has to go beyond industrial economies. Finally, investigating proven technologies in the context of emerging markets is of great practical relevance with regard to climate change mitigation (Kristinsson and Rao, 2008; Hargadon, 2010).

Building on extant literature our research question is: Which factors influence early adopters of an established technology in a highly regulated emerging market? In innovation studies, "diffusion is commonly used to describe the process by which individuals and firms in a society/ economy adopt a new technology" (Hall, 2005). As part of this diffusion process, societies have to balance public and private interests while developing the appropriate institutional framework conditions. To do so, on the one hand social and environmental externalities have to be investigated. In this context, Jefferson (2008) conceptually highlights the need for involving the local communities at the grass root level instead of public top-down decision making, while Aitken (2010) challenges existing positive perceptions by academia and public policy towards wind farm developments. Other scholars contributed to the debate by quantifying the social perspective in Spain and the US respectively (Farizo-Alvarez and Hanley, 2002; Jacquet, 2012). On the other hand but not least, market conditions and private risk-/return structures need to be analysed. In this study we exclude the question of balancing positive and negative externalities of environmentally friendly technologies as analyzed by Sovacool (2009) and Aitken (2010). Hence, the focus of this paper is on the implementation of wind farms through project developers that typically precede private (foreign direct) investment (FDI) studied e.g by Athreye and Cantwell (2007). Hence project developers are early adopters of new technology and they also highly dependent on public policy.

While there is widespread agreement that favourable policies can help adoption and diffusion, there is no consensus what exactly "favourable" means. Even in industrial countries such as Spain it is debated if the successful renewable energy diffusion can

---

7   By barriers to adoption we solely mean market and institutional barriers; focussing on established technologies we can abstract from technological barriers

be attributed either to the Feed-in-Tariff mechanism[8] (Ringel, 2006; Söderholm, 2008) or to other framework conditions (Dinica, 2008; Stenzel and Frenzel, 2008). Moreover, only limited knowledge exists regarding preferences of private sector actors for different policy options for renewable energy, even more so in emerging economies. Recent work extended analysis to wind developers in the EU and USA (Butler and Neuhoff, 2008; Lüthi and Prässler, 2011), yet still do not address adoption in emerging markets. By contributing to closing this gap, we advance the understanding of technology diffusion by exploring generic and renewable energy-specific policy measures on diffusion in emerging markets from a private sector adopters' perspective.

We have chosen wind energy as a technology focus, because it is one of the cheapest renewable energy technologies, hence most suitable for implementation and manufacturing in emerging economies (Lewis and Wiser, 2007; IEA, 2008). Beyond these points the technology is well established in many industrial economies such as Denmark, Spain, Germany (Garud and Karnoe, 2003; Kristinsson and Rao, 2008). However, while large emerging economies such as China and India are currently very active in implementing the technology, smaller emerging markets are still in a nascent stage. Therefore, the latter economies offer a suitable framework for our study. In this context, our research analyzes the relation of the public and the private sector by investigating private decision-making and the role of policy at the early adoption stage. Our main contribution is to identify the role of public policy in triggering diffusion of new technologies in emerging markets; more specifically, we evaluate qualitatively and quantitatively private sector perceptions of various policy measures.

The rest of the paper is structured as follows: First, we analyze theoretical concepts and approaches in the context of energy policies (3.2). This forms the basis for a section on method and data (3.3). The findings of both qualitative and quantitative data that we gathered during the process are presented in detail in chapter 3.4 and discussed in chapter 3.5. The paper concludes with implications in chapter 3.6.

3.2 Conceptual background

Rogers (1995) defined diffusion as "the process by which an innovation is communicated through certain channels over time". The innovation may be a process, an idea, a concept, product, or a set of these, which is newly available to potential adopters. In our case, the innovation is energy infrastructure which is new not because of technological innovation, but on account of deployment in a new region with limited legacy in applying this technology.

---

8   We define Feed-in-Tariff as a government guaranteed fixed price that is payed to the renewable energy plant operator for each kWh that is sold to the national electricity grid (Ringel, 2006).

Based on the work of Metcalfe (1988), the diffusion literature can be divided in two dominant streams: Firstly, those characterizing the mechanisms and patterns of diffusion. This approach looks mainly at the rate and total amount of adoption in a given population within a time period. Secondly, those seeking to understand and to characterize the decision-making structure and process regarding product adoption. This approach focuses on the individual decision-making based on rational choice. Geroski (2000) and McEachern and Hanson (2008) further elaborated on the issue.

The first type of aggregate analysis of the diffusion of technology or more specifically energy infrastructure was very useful to gain a general understanding about FDI in emerging markets Athreye and Cantwell (2007), the impact of market reforms and political uncertainty (Henisz, 2002; Allard et al., 2012) or policy mechanisms in major renewable energy markets – usually industrialized countries. One key observation is the strong impact on the diffusion curves of changes or differences within the national policy framework (Wong, 2005; Späth and Rohracher, 2010; Dewald and Truffer, 2011). Broadly speaking, these studies analyze aggregate market data to find out which policy mechanism would best support market deployment (Jacobsson and Bergek, 2004; Ringel, 2006; Kristinsson and Rao, 2008; Mostafaeipour, 2010; Sovacool, 2010). However, any support mechanism by itself lends a rather narrow perspective that cannot fully explain deployment. For instance, the success of the Spanish market was initially attributed to the implementation of a Feed-In-Tariff (Ringel, 2006; Söderholm, 2008). However, Dinica (2008) and Stenzel and Frenzel (2008) revealed that other factors such as proactive Spanish utilities and public-private partnerships drove the diffusion. Moreover, the main reasoning for a quota system with tradable certificates is to achieve a defined renewable energy target at lower costs compared to a Feed-In-Tariff (Haas et al., 2004). Later studies show that the opposite is true (Lipp, 2007; Bergek and Jacobsson, 2010; IEA, 2008). One can conclude that even in the well-studied context of industrial economies with available and reliable aggregate market data the role of public policy is ambiguous. Therefore, one can neither replicate the study designs in emerging markets due to inferior secondary data availability and quality nor assume transferability of any directionality of above-mentioned results.

The second type of diffusion analysis based on Metcalfe (1988) seems to be more appropriate for the selected focus of analysis. In fact, the micro level analysis of individual decisions to adopt new technologies furthers our understanding of diffusion patterns. Complementary to the aggregate approach, the micro level perspective allows exploring why and when adoption occurs. Here, models and approaches available include a wider array of factors that arguably influence the decision to adopt (McEachern and Hanson, 2008). In the field of energy infrastructure the micro level perspective was first proposed by Wiser and Pickle (1998) who highlighted that many policies are not effective in terms of market growth as they do not meet the needs of investors, which was supported by Enzensberger et al. (2002) who classified the broad term investors on a conceptional level. The investment risk and the related financing costs were identified as key factors that affect the market deployment (Butler and Neuhoff, 2008; Sovacool, 2010; Lüthi and Prässler, 2011). On this basis we analyse influencing factors in the new context of emerging and developing countries.

3.3 Data and methodology

In order to explore how public policies and other factors affect investment decisions of early adopters we have chosen a three-step approach including a mix of qualitative and quantitative techniques (Figure 3.1).

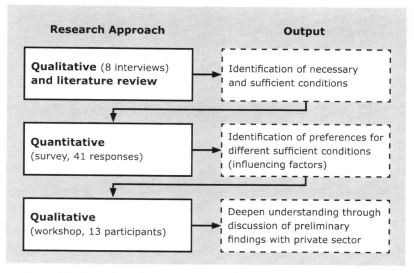

Figure 3.1: Overview of three-step research approach and related output.

3.3.1 Qualitative approach: Identifying the key decision factors

In order to understand the rationale of the decision process of project developers and to identify necessary and sufficient conditions with regard to the decision process, four semi-structured interviews were conducted with German project developers who are active in wind power projects in emerging economies. Key questions were: Which challenges do you face with your current wind energy projects in emerging economies?

How do you choose which emerging market to enter next? The results were further discussed and refined during additional semi-structured interviews with four international experts in the field of renewable energy policy and finance. All interviews were conducted face to face or via telephone and took between 30 and 90 minutes each. The aim of the interviews was to understand the rationale of the decision process of project developers and to identify necessary and sufficient conditions with regard to the decision process.

Necessary conditions are absolutely required for all project developers: Firstly, availability of wind resources; secondly, political stability which means that government including all authorities remains stable in the foreseeable future. Thirdly, financial viability of the project, meaning either a financial support mechanism for wind power or sufficiently high and stable electricity tariffs that ensures an income for the investor; and, lastly, the possibility to access to the grid. If one of these four conditions is not given, then the wind market will stay at a nascent stage no matter how encouraging the other policy mechanisms are.

A second set of sufficient conditions covers items that can have a significant impact on the attractiveness of an emerging markets but not absolutely necessary. These items are further investigated in this paper and belong to either generic or renewable energy specific influencing factors (see also table 3.2).

3.3.2 Quantitative approach: Evaluating the key decision factors

Due to the peculiarities of the cross-national study setting, we decided to use an innovative method called Maximum Difference Scaling that can overcome weaknesses of conventional survey techniques. Butler and Neuhoff (2008) who also study private sector preferences for renewable energy policies in the UK and Germany applied a 5 point Likert scale in their survey. A key finding was that in average the German respondents are 0.6 points more critical than UK respondents. The authors could only speculate if this difference is related to cultural differences or or not.

Maximum Difference Scaling, which was first developed by Finn and Louviere (1992), offers an efficient questioning structure to evaluate preferences for a set of items, in our case a list of 20 (Cohen and Orme, 2004). Respondents are given a fictitious decision scenario (see Appendix) and then they are asked to choose the most attractive item (best) and the least attractive (worst) item from a set of five items. Due to its simple questioning structure, the method prevents scale use bias that could be an issue with other survey techniques as discussed above (Couch and Keniston, 1960; Bachman and O'Malley, 1984; Cohen and Neira, 2003). This is especially relevant in our case as decision-makers with various cultural backgrounds are involved.

As the respondent answers 12 questions with different combinations of items Hierarchical Bayes (HB) estimation is able to extract the relative preference of each respondent and the whole sample regarding all 20 items of the study (see also Table 3.2). HB estimation recognizes that knowing the entire distribution of all individuals' preferences enhances the estimation for each individual. According to Orme (2002) the HB model consists of two levels: firstly, respondents are considered as members of a population of individuals with similar preferences and, secondly, each individual's preferences are calculated according to the respondent's choices within the survey. During 40,000 iterations respondents' preferences are adjusted depending on the amount of variance in the sample. Thereby, HB estimations reflect the optimal mix of individual ratings and sample averages. (Rossi and Allenby, 2003)

3.3.3 Sample selection for and sample characteristics of the survey

We followed a two-step process to identify those decision-makers that have the necessary experience to rank the list of items. Firstly, we identified companies that act as project developers for wind energy in at least two countries including at least one emerging market. After conducting a comprehensive research on the web and checking the membership directories' of international and national wind industry associations88 companies were found. Secondly, we identified up to three decision-makers in each company who each had already gained personal experience with wind energy in emerging economies. A total of 158 persons were identified.

The survey was implemented between June and August 2010. Before launching the survey, a pre-test was conducted with five experts in order to validate the measurement and refine the survey. Then, potential respondents were invited individually, by email or by social networks such as LinkedIn and Xing. People who did not reply within four weeks were contacted again. To optimize the accuracy of responses and to limit the impact of self-assessment, we guaranteed that all information would remain completely confidential, we promised to share the final results of the study with respondents, and we agreed to distribute a personalized feedback document (Huber and Power, 1985). This process allowed us to gather 41 completed questionnaires by persons from 36 different companies, corresponding to an effective response rate of 26 % in terms of individuals and 41 % in terms of companies of the total population. Seven additional responses were returned incomplete; therefore, they were not taken into account for further analysis. Regarding the personal background, the respondents gained work experience in a large number of emerging markets (Figure 3.2). Economies that were named only once or twice are summed up per continent. Other descriptive statistics are presented in Table 3.1.

Table 3.1: Descriptive statistics of the research sample

	N	%
Location of headquarter of the company		
Germany	14	34
EU ex. Germany	15	37
Non EU	8	19
No answer	4	10
Strategic regional focus of the company[1]		
Europe, US	29	72
BRIC	22	55
small emerging economies	19	49
Strategic technology focus of the company[1]		
Wind power (on-shore)	40	98
Wind power (off-shore)	8	20
Other renewable energy technologies	18	44
Other non-renewable energy technologies	5	12
Coverage of the value chain by the company[1]		
Acquisition and preparation	26	63
Planning and technical feasibility	34	83
Development and approvals	38	93
Financing	20	49
Construction	19	46
Operation	26	63
Department of the respondent		
Top management	19	46
Business development	12	29
Project management	9	22
Finance	1	3
Professional wind energy experience of the respondent		
Up to 5 years	28	68
5 to 10 years	7	17
10 to 15 years	0	0
more than 15 years	6	15

[1] multiple answers possible

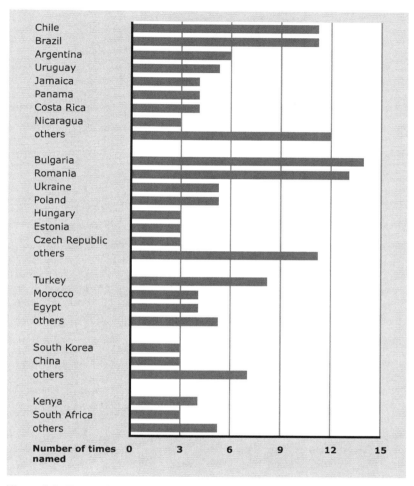

Figure 3.2: Economies in which study participants gained professional experience (multiple answers possible)

3.3.4 Qualitative approach: Discussing the key results during a workshop

In order to refine and corroborate our understanding of the interviews and the survey data a workshop with industry experts was conducted. The participants of the workshop had a background in finance and assurance (3), project development (2), manufacturing (1) and consulting (4). In addition, three national and international wind energy industry associations were present. During the workshop the research team presented the intermediate results and discussed each item. In order to minimize the impact of one expert on the overall results we decided not to invite any interview partner for the workshop.

3.4 Findings

A summary of the qualitative findings as well as of the preferences based on the 41 completed surveys is shown in Table 3.2. Since the overall theme is related to policy support mechanisms, factors are hardly assessed negatively by default; other factors were captured by inversely worded questions in order to identify the varying degrees of only positive statements. In case all 20 items were evaluated equally important, the preference for each item would be 5 % (100 % divided by 20).

The research team derived the evaluation based on the interviewees' and the workshop participants' points of view. Therefore, we decided to apply a simple scale (0 = neutral, + = support and ++ = strong support) that indicates the attractiveness of each framework condition aggregating both interviews and workshop data separately. These indicators aim to give a broad overview of the qualitative results to better interpret the context of the quantitative survey.

This chapter is divided in generic and specific influencing factors and includes all findings using both qualitative and quantitative data, i.e. interviews, survey and the workshop. One or several items are grouped into factors in order to draw propositions for the preference of international private sector decision makers (see Figure 3.3). For further interpretation it is important to bear in mind that all 20 items are identified to be important for a project developer. Therefore, in some countries it might happen that an item of low overall importance becomes critical for the success of the project.

Table 3.2: Overview of the results of qualitative and quantitative measures (propositions (P), interviews (I), survey (S), workshop (W))

P	Item	I[1]	S (%)	W[1]
1	Good legal security (contracts are easily enforceable)	+	8,586	++
	Low risk of unforeseen policy changes for renewable energy	++	7,170	++
2	High degree of transparency in the approval process	++	8,051	++
	Maximum duration of the administrative approval procedure: 18 months	+	6,836	++
	One-stop-shop for all necessary approvals	+	4,966	0
3	Availability of attractive financing by development banks	+	6,470	+
4	No inflation risk (tariff adjusted to the inflation rate)	+	6,635	+
	Reduced currency risk, tariff adjusted to exchange rate of the US$	+	3,461	+
5	Feed-in-tariff with fixed prices for 20 years	++	13,144	++
	Investment based support (accelerated depreciation and investment tax credit)	+	3,565	0
	Intern. tender for a framework contract for 200 MW incl. PPA for 20 years	+	1,624	0
	Quota-system with Tradable Green Certificates	+	1,461	0
6	Grid access and priority dispatch are guaranteed	++	10,814	++
	The off-taker has a good credit rating	+	4,101	0
	It is clearly defined who pays for which part of the grid connection	+	2,159	0
	Grid access and dispatch are regulated but not guaranteed	+	1,492	+
7	Availability of attractive financing by local banks	+	4,717	+
	A local wind developer is ready to set up a joint venture with you	+	2,254	0
	Bonus payment of 1$ct per kWh if major components are procured locally	+	1,556	0
	Trained local technicians in the field of wind energy are available	+	0,938	+

[1] 0 = neutral, + = preference, ++ = strong preference

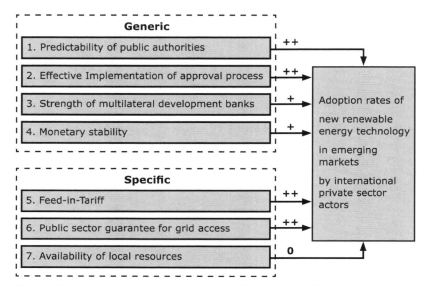

Figure 3.3: Overview of propositions derived from the qualitative and quantitative data

3.4.1 Generic influencing factors

The first set of influencing factors deals with the broader framework for infrastructure investments in emerging markets. This considers not only the way in which government is run but also development cooperation and monetary aspects. As mentioned above, these are all known for their positive contribution to FDI in general; however our aim was to identify a prioritization in the context of renewable energies, more specifically wind.

- Predictability of public authorities

The need for more predictability and stability is highlighted by the item "good legal security" (8.586 %). In this context one project developer stated:

"Our priorities for a market entry decision are clear: high legal security and low level of corruption." (Project developer, interview).

Moreover, predictability can also be measured by the risk of unforeseen policy changes (7.170 %). In the past some emerging economies decided to set up very attractive framework conditions for renewable energy in their nascent local markets. These decisions immediately attracted many international project developers and investors to start their operations in these economies. Surprisingly, e.g. in Eastern European economies, many of them failed (Noethlichs, 2011). Fast and necessary changes of framework conditions were necessary which often results in strong effects on financial viability of projects. This ultimately leads to low trust of investors who favour long-term stability and predictability (Table 3.2, IEA (2008)).

"We need to go forward step by step and slow down the gold rush that can be created by an ambitious emerging market." (Financial institute, workshop).

To summarize, predictability of public authorities is strongly positively related to adoption rates.

- Effective implementation of the approval process

Building again on the example of some Eastern European countries, many interview partners identified the lack of definition regarding approval processes as a major barrier. More specifically, emerging economies are evaluated to be very attractive if the approval process is transparent (8.051 %) and well-defined resulting in low durations of the approval process (6.836 %). This standpoint was clearly confirmed by the empirical study and the workshop participants.

"For developing our new international strategy we first looked at the corruption index of all countries" (Project developer, interview).

"You never know how regulations are implemented–although they are approved by the government!" (Turbine manufacturer, workshop).

Regarding the approval process several experts mentioned a one-stop-shop for necessary approvals could be a very attractive measure. They clarified their statements as based on experiences in industrialized economies. Interestingly, our empirical findings regarding emerging markets show no clear preference regarding a one-stop-shop (4.966 %). In the workshop it became clear that this simplification of the approval process bears several risks. For example, less transparency as well as less knowledge of and influence on the approval process. Although this approach is built on good intentions, workshop participants see a high risk that it ends up in an additional administrative barrier. In summary, this study does not support the implementation of a one-stop-shop in emerging markets.

"Based on our experience in Eastern European markets, I believe a one-stop- shop will most likely lack the political power to truly accelerate the approval process." (Wind energy association during the workshop).

In combination, effective implementation of the approval process is strongly positively related to adoption rates

- Strength of multilateral development banks

The international focus of many project developers might also facilitate acquisition of international public support, e.g. through multilateral development banks. During interviews it became evident that attractive financing conditions from development banks are relevant items for market entry of international project developers although this always results in significant administrative effort. While the survey shows only slight preference for this item compared to all other items, workshop participants conclude that especially during the first wind farm projects, development banks play a role. Therefore, the strength of multilateral development banks is positively related to adoption rates

- Monetary stability

According to our interview partners a policy framework that covers inflation risk can be a success factor for financial viability of a project in some countries. One reason is that initial investment for constructing a wind park is often made in Euro or US Dollar while income generated during the operation phase is received in local currency. If inflation increases more than expected during the operation phase investors face a loss. Therefore, this item is rated as attractive in the survey (6.635 %). Workshop participants confirm this and underline that an inflation-adjusted tariff is relevant in some emerging economies. Even if risk is only partly transferred, it has a relevant impact on investment decisions. Further, workshop participants discussed why reducing inflation risk (6.635 %) is rated more attractive than reducing currency risk (3.461 %). It turned out that hedging currency risk seems to be easier through financial markets because initial investment and the repayment of debt are well-known at the beginning of the project. To conclude, monetary stability is positively related to adoption rates.

3.4.2 Specific influencing factors

Beyond the generic influencing factors, there are also certain factors specific to renewable energy infrastructure investments.

- Feed-In-Tariffs and other support mechanisms

Quantitative results show that Feed-In-Tariffs are rated significantly more positive (13.144 %) than all other support mechanisms (3.565 %, 1.624 %, 1.461 %, respectively). During interviews it became clear that many project developers gained very positive experience with Feed-In-Tariffs. A fixed price per kWh guaranteed by government considerably reduces investor risk and creates a stable income. This risk reduction helps early stage project developers going through initial project phases and reduces cost for debt (interest rate) and equity at a later stage of project development. Our quantitative study and workshop participants fully support this positive evaluation.

"For us, a Feed-In-Tariff is a must-have, the only way around would be a market that has already many successful wind farms." (Project developer, interview).

During interviews two types of investment-based support mechanisms were highlighted: Firstly, accelerated depreciation of the investment by which large companies can reduce their tax burden. However, small companies or companies that are not profitable yet do not benefit from this support. Secondly, investment can be supported by grants. However, such grants are often paid only after commissioning of the wind park. This means that project developers bear additional risk during construction phase, as grants may not be paid as initially planned. This might partly explain why project developers evaluate this support mechanism as little attractive in the survey (3,565 %); workshop participants share this view.

When international tenders are announced, there are several challenges for project developers according to our interview partners. Due to complexity and length of the tendering procedure, this mechanism is only applied for large power projects. Often large project size goes beyond the capabilities of small and medium project developers. Additionally, the mechanism does not allow for continuous market development. Oftentimes, it is not transparent when and under which conditions government issues a second or third tender.

"A tender is always difficult for a project developer–especially, if it is about a one-time project." (Wind energy association, workshop)

Hence, it becomes clear why project developers do not rate this mechanism as an attractive framework condition in the survey (1.624 %). In the workshop another risk was mentioned: Compared to other support mechanisms, tendering often also opens the door for corruption.

A fourth relevant support mechanism according to our interview partners is a quota system with tradable green certificates (1.461 %). This mechanism was initially developed in order to increase price competition and focus on the most mature technologies at the best sites within a country. Unfortunately, experience contradicts originally good intentions of this mechanism. Qualitative data from both interviews and workshop confirm that quota systems are mainly an additional income risk. This risk requires financing institutions and equity investors to ask for an additional premium, which significantly increases overall project costs.

"A Quota System always requires more equity compared to a Feed-In-Tariff– the leverage would be 65/35 rather than to 75/25." (Financial institution, interview).

As the qualitative data from interviews and workshop fully support survey results we conclude that feed-In-tariffs are the only support mechanism strongly positively related to adoption rates.

- Public sector guarantee for grid access

During interviews it turned out that well-defined regulation regarding grid connection and power dispatch is far better than no special regulation for renewable energy. However, the difference between guaranteed grid access and well-defined regulation was never clear. Therefore, the research team decided to include both items in the study. Interestingly, the survey revealed that a guarantee belongs to the most attractive policy measures (10.814 %) while a regulation (without guarantee) belongs to the least attractive (1.492 %). Based on explanations of workshop participants, guaranteed grid access is a very strong political signal for implementation of renewable energy projects.

Beyond regulations that directly address grid access, financing wind parks in emerging markets also depends on the credit rating of the off-taker who is usually a public utility. If the off-taker is not sufficiently credit-worthy, other government guarantees regarding the Power Purchase Agreement are required. The survey revealed that project developers do not have a specific preference for this item (4.101 %). In the workshop it turned out to be relevant in only a few developing economies.

During interviews it turned out that it is often undefined who has to pay for grid access of a wind farm and potential grid extensions. However, the survey revealed a low preference of this item (2.159 %) compared to all other items, as long as access will be provided. During the workshop the participants underlined that this matters only in some cases depending on country and site. Therefore, we conclude that public sector guarantee for grid access is strongly positively related to adoption rates

• Availability of local resources

Interviews with project developers revealed that market entry in emerging economies often starts with a joint venture with a local company. The main reason is the need for personal contacts to local policy makers in order to facilitate and streamline the approval process.

"In order to enter a new emerging market the key is to have local partners with good relations to policy makers." (Project developer, interview).

However, in the empirical study this item was rated as little attractive in comparison with all other items (2.254 %). The research team found two different reasons for this rating: on the one hand, it is possible that this item was overrated during the interviews. Therefore, the empirical study corrects the first impression. On the other hand, it is possible that there are enough potential joint venture partners. Therefore, it is not difficult to find a suitable local partner for an international project developer. It was not possible to finally answer this question during our workshop.

Regarding local and potentially government owned banks, our interview partners highlight that including them means to have a supporter who is well connected with public authorities. This helps to assure the long-term project success.

"It is a huge advantage to have a local or even government owned bank as a financial partner as this helps to minimize the risk of expropriation and to increase pressure on policy makers." (Project developer, interview).

However, workshop participants underline that local banks often lack experience and knowledge to finance wind farms. Therefore, project developers or international donor agencies often have to strengthen these capacities at the local bank.

In order to agree on financial support mechanisms policy makers often argue for local value and job creation. Local manufacturing can either be enforced through official regulation or facilitated through voluntary mechanisms. During interviews several project developers were very critical regarding regulations that force them to procure parts of the wind farm locally. However, it was not clear how the private sector evaluates an incentive mechanism giving a financial bonus if major components are procured locally. The survey shows that such regulation is not attractive for project developers as compared with the other 19 items (1.556 %). In the workshop with industry experts it became very clear that local content requirement is challenging, especially in nascent markets. As there are only few potential suppliers in a country that has just started to use its wind resources these companies are tempted to charge monopoly prices for their technology. However, there is a clear opportunity for local companies to become involved. In summary, the study clearly shows that on average local content incentives are not attractive for international project developers.

"Our clients often ask for local production which is not financially viable for us to do it ourselves. In some cases, we source tower elements locally – if the know-how is available." (Wind turbine manufacturer, workshop)

To conclude, availability of local resources has limited effects on adoption rates.

3.4.3 Robustness check

A key question of all surveys is whether the stated preferences match with real-life decisions. In order to answer this question study participants revealed how many wind projects they finished successfully under each of the four different support mechanisms (see Proposition 5). As the correlation between real-life investment decisions and revealed preference for the four support mechanisms (see also Table 3.2) is high (Pearson's correlation 0.94), we argue that the method of Maximum Difference Scaling is able to derive real preferences of decision-makers in the context of investments in energy infrastructure. In addition, we conclude that this high correlation between the preference for a support mechanism and the number of implemented projects under the given support mechanism corroborates the overall quality of all revealed preferences.

3.5 Discussion

Most existing diffusion research analyzes aggregate phenomena and concludes how effective different framework conditions are. This approach is effective in established or more mature markets. Hence individual decisions and preferences of the private sector and other stakeholders can be measured indirectly by analyzing historical market development with available and reliable data sources. Our study takes a different lens by directly measuring individual private sector decisions in the early adoption phase. We applied this approach to emerging or nascent renewable energy markets in which the relevance of a disaggregate approach is even more striking, at least partly due to (secondary) data limitations. We contribute on one hand to the literature on diffusion of technological innovations in emerging markets such as McEachern and

Hanson (2008), Kristinsson and Rao (2008) and Lee (2009). On the other hand we enhance the ongoing debate among both academics and policy makers about efficient and effective public policies to foster "rapid and widespread adoption of alternative energy technologies" (Mowery et al., 2010). Thereby, we expand the geographical reach of the literature on renewable energy investment decisions beyond the earlier focus on EU and USA (Butler and Neuhoff, 2008; Lüthi and Prässler, 2011) and show that an implicit assumption on policy formulation in industrialized countries does not suffice for emerging and developing countries due to "institutional voids" (Khanna and Palepu, 1997).

With regard to renewable energy-specific factors our findings corroborate quantitatively prior qualitative work on support mechanisms, such as Feed-in-Tariffs, in Asian economies (Sovacool, 2010). Thereby, our results of individual preferences in the context of emerging economies also support findings derived from analysis of aggregate market data in industrialized countries (Lipp, 2007; Bergek and Jacobsson, 2010; IEA, 2008). In fact, our qualitative results underline that in emerging markets Feed-in-Tariffs combined with guaranteed grid access are even more important than in industrialized countries. Both mechanisms considerably reduce comparatively high investment risk, which is typical for emerging countries (Khanna and Palepu, 1997; Henisz, 2002).

Our results show that in emerging markets – in addition to technology-specific factors – generic influencing factors such as transparency and legal security for international private sector organisations must be considered. We add to the (renewable) energy policy literature, which focuses on policy formulation, by emphasizing these implementation factors for emerging markets. Predictability of government decisions or low risk of unforeseen policy changes is evaluated to be very important for investments in wind farms which is in line with existing literature such as Henisz (2002) who performed a two-century long historical analysis regarding infrastructure investment. Beyond defining appropriate framework conditions, effective implementation emerged as a major influencing factor. Clearly defining and implementing the entire approval process includes not only national and local governments but also utilities and grid operators. We conclude that improving the implementation can potentially have a tremendous effect on international private sector actors.

In addition, our data reveal mixed results regarding local companies and co-investors that can possibly reduce risks during project development and operational phase. Although local actors have an extended local network and understand how to deal with not fully implemented policies and other institutional hurdles, the findings on local resources reveal a limited effect on adoption decisions. In line with Lund (2009), workshop participants argue that local companies typically start manufacturing tower elements and at a later stage potentially extend their contribution along the value chain. While this path could be tremendously facilitated by political stability and attractive investment conditions or pro-business market reforms (Allard et al., 2012) our study re-vealed that local content requirements – a central approach of policy makers in emerg-ing markets – can cause more harm than benefit even if financial incentives are in place.

In the broader debate on effective regulatory framework conditions for developing and investing in renewable energy, one suggestion from our qualitative findings is that decision-makers from the private sector should be involved in early stages of policy making. In line with Wiser and Pickle (1998), Enzensberger et al. (2002), Gross et al. (2009) and IEA (2008), one key recommendation for policy makers is to focus on a suitable balance of risk and return for the investor. Lower risks for investors conse-quently lead to lower financing costs and therefore lower overall costs for the project (de Jager and Rathmann, 2008). More precisely, our qualitative data show that in-stead of one or few bold improvements of framework conditions small, yet predictable, regular steps to constantly improve investment conditions are perceived as leading to best results in terms of market growth, investor confidence, creation of a local or do-mestic value chain and local employment (Lewis and Wiser, 2007; Mostafaeipour, 2010).

This explorative study has several limitations that could lead to fruitful further research. Firstly, the validity of the quantitative results is limited to the given scenario. There-fore, using data from the survey in order to draw conclusions for much smaller or larger wind parks, as well as for other renewable energy technologies or energy infrastructure projects, is only possible to some extent. Secondly, as the level of analysis of our 20 items is rather high it does not allow us to investigate variations or interactions at a more detailed level. Such variations, in turn, can affect the overall attractiveness of the mechanism (Ringel, 2006; Dinica, 2006; Couture and Gagnon, 2010). Thirdly, as the sample size is comparatively small we could not analyze different sub-groups of our sample. Exploring regional variations would lead to more nuanced findings and im-plications. Fourthly, our main argument is to improve the public decision making pro-cess by providing methods to capture private sector perceptions. For this purpose we deployed private sector feedback on various scenarios reflecting different framework conditions. We do not investigate social and other external costs of these scenarios. Thus, our results imply individual private sector preferences rather than full cost anal-ysis (including external costs) or welfare-theoretic economic optima. Building upon existing methods, approaches and perspectives that were developed and applied in the industrialized country context such as Farizo-Alvarez and Hanley (2002), Jefferson (2008) and Jacquet (2012) further research could explore potential public and private benefits and costs. This could include aspects as diverse as job creation and gains from renting land to wind farm owners and negative effects such as noise intrusion and impact on property values in the context of wind energy.

3.6 Conclusion

So far only limited knowledge exists regarding preferences of private sector actors for different policy options for renewable energy and the role of institutional context, even more so in emerging economies (Popp, 2010; Kemp and Oltra, 2011). While there is widespread agreement that favourable policies can help adoption and diffusion, there is no consensus what exactly "favourable" means. Therefore, policy makers oftentimes develop and adapt regulations without evidence on the effect on private investors. This study is a first attempt to close this gap and explores among the plethora of policy instruments those that are deemed most effective for technology diffusion from a private sector perspective. It sheds light on the preferences of project developers in emerging markets paving the way for (international) investors.

On the one hand, we could show that a Feed-In-Tariff with fixed prices for 20 years and Grid access and priority dispatch guaranteed are by far the most attractive framework conditions for project developers. Diffusion of renewable energy technology is fostered by these policy measures because developer attractiveness is increased as an important condition for investments. On the other hand, one-stop shops for the approval process, which is proven to be effective in industrial countries, is not seen as appropriate in emerging markets. Instead, our qualitative findings suggest that policy makers should increase the attractiveness of their renewable energy market by firstly, focussing on effective implementation of policies which includes also capacity building and secondly, improving the policy framework in small but regular and predictable steps which also allow for the growth of a domestic industry.

The categorization of these measures in renewable energy technology-specific and generic enabled us to show systematic differences against findings from extant literature in industrialized countries. The implicit assumption on policy formulation does not hold in emerging markets contexts due to "institutional voids" (Khanna and Palepu, 1997). While industrialized countries "only" need to set up a support mechanism, emerging markets must improve predictability and policy implementation, too (Figure 3.3). This task for public policy is implicitly assumed to exist in industrialized countries, which limits the transferability of extant research on energy policy and technology diffusion.

To conclude, this explorative study is a foundation for further investigations on a country specific level and a starting point for discussions among public and private stakeholders on the national and international level rather than a blueprint for emerging market policy frameworks for renewable energy technology diffusion. More specifically, the role of different institutional frameworks across emerging markets warrants deeper scrutiny. Although we are aware that our approach does not represent the full complexity of investment preferences, we believe that the combination of qualitative and quantitative research allows drawing valid conclusions also for the ongoing debate regarding institutional investors or regarding national and international climate change measures such as the UNFCCC Green Climate Fund and National Mitigation Action Plans (NAMAs). These public measures can be used to fully or partly finance, for example, Feed-In-Tariffs in developing and emerging economies or to provide any other form of guarantee for private investors.

4 Exploring the link between products and services in low-income markets – Evidence from solar home systems

Christian Friebe, Paschen von Flotow, Florian A. Täube
Energy Policy 52 (2013), 760-769[9]

Abstract

One of the key challenges of energy access in emerging markets and developing countries is how to reach households and communities that are unlikely to get a grid connection in the long term or those that are connected to the grid but suffer from regular blackouts or low voltage. By surveying entrepreneurs selling Solar Home Systems (SHSs) on a commercial basis in emerging and developing countries, this study is one of the first attempts to quantify the key elements of four potential Product Service Systems (PSSs): Cash, Credit, Leasing and Fee-for-Service. Whereas the Fee-for-Service approach was found to be suitable only under certain conditions, all PSSs share two key elements for successful market deployment: one or more years of maintenance, and customer support in financing these customers' new asset. Moreover, it appears that private sector companies are in principle able to deliver SHSs to households with incomes greater than USD 1000 per year. The implications for policy makers and development aid agencies are, first, to include maintenance services into public programmes or public–private partnerships and, second, to explicitly consider financial risks for entrepreneurs (e.g., customer commitment and repayment conditions).

---

9   Presenting and discussing earlier versions of the paper at the "oikos UNDP Young Scholars Development Academy 2011" in Bangalore, the "Business of Social and Environmental Innovation 2011" in Cape Town, the "ETH PhD Academy 2012" in Appenzell and the "Academy of Management 2012" in Boston helped tremendously in developping the paper.

4.1 Introduction

Whereas many products and services have the potential to increase the standard of living or to stimulate economic activities in emerging and developing countries (London et al., 2010), researchers as well as practitioners and policy makers agree that energy access is not only an objective in itself but also an enabling technology that leads to many other important innovations (Gustavsson, 2007; Rao et al., 2009). According to the IEA et al. (2010), 1.4 billion people still have no access to basic electricity services. Although the potential market size for energy services is substantial (USD 433 billion, according to Hammond et al. (2007)), thus far, only a few companies are commercially successful in this market. The missing link between huge demand and limited supply is a sign of market failure that deserves the attention of policy makers.

Given the challenges of a limited public budget and broader commitments such as the Millennium Development Goals (MDG), many national and international policy makers aim to accelerate the diffusion, promotion, and development of off-grid power supplies on a commercial basis. However, many policy programmes in the past did not meet the expectations of national policy makers and donor agencies (Acker and Kammen, 1996; Sebitosi and Pillay, 2005; Wamukonya, 2007; Vleuten et al., 2007). Therefore, it is necessary to better understand how successful entrepreneurs and their energy service companies design products and related services to address customers' needs and expectations (see also Fig. 4.1). This paper focuses on the Solar Home System (SHS), a combination of a photovoltaic module, a battery, a charge controller and efficient light bulbs that serves individual households, as a key technology that can be applied in many regions in emerging and developing countries. According to the World Bank, the largest market potential in SHS reflects Leasing and Fee-for-Service customers (up to 70 %), e.g., the largest and poorest sections of society that cannot afford to buy an SHS outright (Worldbank, 2008). Whereas some case studies focus on companies that offer Fee-for-Service as part of a public–private partnership (e.g., Lemaire (2009)), it is not clear how purely commercial companies are able to respond to this substantial demand and how they adjust their product and service offerings. This uncertainty calls for an inquiry into the motivations and preferences of entrepreneurs who deploy the large potential for SHS in low-income markets and how they do so.

A better understanding of this private sector perspective enables policy makers to first improve framework conditions for the SHS sector and its local entrepreneurs and to use limited public funding more effectively and efficiently. This is especially relevant as the entrepreneur directly interacts with consumers and their view and expectation on different product designs and related services. Therefore, the guiding question of this paper is as follows: "Which combinations of products and services are reasonable businesses for entrepreneurs in the SHS sector?" This research question is addressed by expert interviews and a quantitative explorative survey using a recently developed variant of the conjoint method (see Shepherd and Zacharakis (1999); Patzelt and Shepherd (2009); Loock (2012) for other variations of the conjoint method).

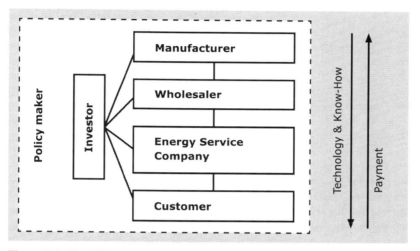

Figure 4.1: Simplified value chain for delivering SHSs to customers in emerging and developing countries

In this way, this research complements and refines existing, mostly qualitative research on low-income energy markets. So far, researchers have conducted, on the one hand, surveys and interviews with end-users (McEachern and Hanson, 2008; Ndzibah, 2010; Abdullah and Mariel, 2010; Sovacool et al., 2011) or qualitative case studies of individual companies (Lemaire, 2009; Mukherji and Jose, 2010). These studies show a rich variety of contextual factors, such as knowledge, awareness, finance and social aspects. On the other hand, there are macro- level case studies that focus on one or several countries or technologies (Karekezi and Kithyoma, 2002; Hammond et al., 2007; Ketlogetswe and Mothudi, 2009; IEA, 2010; Chaurey and Kandpal, 2010; Rebane and Braham, 2011). These studies gen- erally produce recommendations for policy makers at the national or international level.

Examining existing studies with the lens of diffusion of innovation (Metcalfe, 1988; Sarkar, 1998; MacVaugh and Schiavone, 2010) reveals that only a few studies (for example, McEachern and Hanson (2008)) explicitly address both the microlevel of individual adoption decisions and the macro-level of countries and technolo- gies. Our approach is to investigate patterns of companies across a variety of countries – the macroperspective – by conducting a quantitative survey among entrepreneurs – the microperspective. In this way, our study contributes to the on-going research debate that questions and refines our current understanding of low-income markets (see, for example, Prahalad (2012); Seelos and Mair (2007); Anderson et al. (2010)).

The paper proceeds as follows: first, the research context, Energy Services and Product Service Systems, is introduced. Next, Product Service Systems are specified for the context of SHSs in low-income markets, which is followed by a discussion of the method and data. Subsequently, the findings are presented and discussed. We conclude with implications for policy and future research areas.

4.2 Research context

4.2.1 Access to modern energy services

While practically every public stakeholder in developing as well as industrialised countries agrees on the goal of achieving "universal" modern energy access, the definition remains under debate. Three issues require further elaboration, and we surveyed the extant literature accordingly:

First, from an engineering and a business perspective, different types of infrastructure are required to allow for electricity access. In rural areas, a mix of grid expansion, mini-grids and off-grid energy infrastructure is most suitable economically and technically. The optimal mix of different technologies depends largely on local demand, the natural resources available and the policy framework (Kaundinya et al., 2009; Levin and Thomas, 2012). The first approaches to compare different energy infrastructure types and their socio-economic consequences are developed (Wamukonya and Davis, 2001). However, in this paper, the focus is placed on the off-grid power supply because this is the most suitable starting point for large parts of the rural population that currently have no access to electricity services.

Second, the question of what modern energy exactly is must be considered. Brew-Hammond (2010) argues that the term modern energy is used to distinguish between traditional forms of technology (e.g., wood) and new technologies (e.g., electricity services). In this paper, the focus is placed on renewable energy technologies and, in particular, SHSs. Although other renewable energy technologies and even conventional fuels are relevant in some cases, solar energy technologies became commercially viable in many applications (Casillas and Kammen, 2011), especially due to their low maintenance requirements and the possibility for flexible system designs.

Third, what exactly does "energy access" mean? The definitions available often remain vague. One example is "access to clean, reliable and affordable energy services for cooking and heating, lighting, communications and productive uses" (AGECC, 2010). More specifically, other definitions state that the term access refers to a household's ability to obtain a modern energy service (Ranjit and O'Sullivan, 2002; Komatsu et al., 2011). In this case, access is a function of two interrelated factors: availability and affordability (Prahalad, 2012; Nakata and Weidner, 2012). In this paper, affordability is defined in relation to the level of household income.

Beyond the question of definition, it is clear that electricity services have a tremendous positive effect on the lives of people previously not using electricity. Major benefits include savings in energy costs and the improvement of living conditions. Existing studies measure these benefits, often by conducting field studies (Gustavsson and El-legard, 2004; Gustavsson, 2007; Obeng et al., 2008). Beyond pure economic factors, other aspects, such as entertainment, the number of children and mobile phones, as well as safety considerations (replacing kerosene lighting) influence buying decisions (Wamukonya and Davis, 2001; Wijayatunga and Attalage, 2005; Komatsu et al., 2011). However, many issues remain to be solved (see Table 4.1).

Table 4.1: Barriers to the use of SHSs in developing countries, compiled by the author based on Urmee et al. (2009), Sovacool et al. (2011)

Category	Issue
Financial	Availability of capital
	High capital cost / high interest rates
	Lack of financing for the programme
	Lack of access to credit for the consumer
	No link to income generation
Technical	Limited product availability
	Logistical problems
	Improper maintenance
	Technical limitations (efficiency, capacity)
Policy	Lack of policy and legal framework
	Improper use of subsidies
	Donor dependency / donor driven
Implementation	Lack of institutional capacity
	Lack of technical knowledge
	Lack of private sector involvement
	Lack of involvement of local stakeholders
Social	Misperception regarding the technology
	Missing link to existing social structures and values

To address the existing barriers to technology diffusion, policy makers and international donors developed public subsidy schemes and support programmes that are well intended but in many cases not sustainable. In fact, some scholars argue that donor efforts are often either ineffective or even undermined existing private sector initiatives, especially those of small and medium-sized local companies (Acker and Kammen, 1996; Sebitosi and Pillay, 2005; Wamukonya, 2007; Vleuten et al., 2007).

Due to the setbacks in the past, policy frameworks, sometimes combined with funding from international donors or develop- ment banks, shifted towards private sector development (Martinot et al., 2002). The aim of this shift was to avoid past mistakes and to facilitate market deployment for accessing electricity services. Development banks that act proactively in this context may have a positive and sustainable impact on the private sector (George and Prabhu, 2003). Depending on the local conditions and on the existing regulatory framework, a blend of different mechanisms is required. This blend includes capacity- building measures as well as tailored subsidy schemes. To design effective support policies such as cash grants or favourable refinancing conditions, it is key to understand the challenges from a private sector perspective.

4.2.2 Product service system – Towards further integration

To access low-income markets, suitable products must be designed. However, in most cases, doing so is not enough. Additional services – beyond conventional after-sales-services – such as consumer training, installation, maintenance and finance must be provided to create profitable companies (Worldbank, 2008). With regard to the given case, Product Service System (PSS), a concept that was originally developed for mature markets in industrialised countries, seems to be useful (Mont, 2002). More specifically, a PSS can be defined as "as a system of products, services, supporting networks and infrastructure that is designed to be: competitive, satisfy customer needs and have a lower environmental impact than traditional business models" (Mont, 2002). The aim of PSS is essentially to develop integrated solutions that also help to protect the environment. This relates nicely with business models that focus on SHS as in most cases the product not only includes technical components but relates also with advisory and maintenance services (IEA-PVPS, 2003; Krause and Nordström, 2004). In addition, environmental protection features strongly in SHS in the context of developing countries, as this product fosters green decentralised energy access instead of small diesel engines for electricity supply (IEA, 2010).

Combining products and services to different degrees has implications for both consumers and companies. On the one hand, such a combination requires consumers to shift from buying and owning products to buying integrated system solu- tions, which often effectively requires better consumer education and involvement (Mont, 2002). On the other hand, companies require a higher level of responsibility for the product as well as early interaction with consumers to achieve an optimal design with minimal environmental impact.

Whereas a general trend towards systems integration can be observed in many different industries, especially in mature markets (Hypko et al., 2010; Loock, 2012), the concept of PSSs focuses explicitly on realising positive environmental effects. It is very important to realise that not only pure technical questions but also other key influencing factors, such as public policies or institutional aspects as well as socio-cultural aspects, are relevant for the successful implementation of PSSs (Mont and Lindhqvist, 2003; Tukker and Tischner, 2006). This paper explores some of the most relevant aspects for the context of SHSs in low-income markets.

4.3 A PSS for SHSs in low-income markets

In the context of no access to the electricity grid, SHSs are more environmentally friendly than conventional off-grid power supply systems that are typically based on fossil fuels. To identify a relevant PSS for SHSs, the work of several researchers has been consolidated at a conceptual level (IEA-PVPS, 2003; Krause and Nordström, 2004; Chaurey and Kandpal, 2010). The four different PSSs are presented below and summarised in Table 4.2.

4.3.1 Cash

The consumer pays for and receives the SHS, which is installed by the consumer himself or by the company. On completion, ownership is transferred to the consumer. The major benefits for the company include low capital requirements and minimal requirements regarding service infrastructure, whereas major risks include a loss of reputation due to system failure that is related either to low-quality components or to insufficient consumer education, poor system design, sizing and performance. Because this PSS is the most capital-intensive scheme for poten- tial consumers, the use of cash transactions is expected to occur only above a given threshold income.

4.3.2 Credit

The consumer receives an SHS and pays regular instalments plus possibly a down payment. The loan may be provided by the company that sells the products or by a financial institution. This loan requires either a financially strong company or an equally strong partnership between the company and a financial institution. Both the co-operation and the financial involvement of the company result, in general, in better consumer training as well as more reliable maintenance services. Given that this PSS may benefit from a large network of micro-finance institutions in emerging and developing countries, it is expected that entrepreneurs prefer it in the context of limited financial resources on the part of the consumer

4.3.3 Leasing

The consumer is allowed to use the SHS and pays regular instalments. Initially, the company owns the system. Later, once the system is fully paid for by the consumer, the ownership is transferred. Similar to a credit system, this mechanism often involves a financial institution for refinancing. Due to the transfer of ownership only at the end of the repayment period, this PSS is expected to require advanced sales and maintenance services to be feasible.

4.3.4 Fee-for-Service

The consumer is allowed to use an SHS that is owned by the company. The consumer pays either a fixed fee for the system uptime or a variable fee depending on the kWh used. In both cases, it is in both the company's and any involved financial institution's interest to keep the SHS up and running in the long term. Maintaining the system includes the proper training of employees regarding a correct installation and maintenance as well as the proper training of consumers regarding the use and limits of the SHS. Again, this PSS is only feasible if advanced services are included. Because the ownership is never transferred to the consumer, this PSS includes considerable risks for entrepreneurs.

Table 4.2: Definition of the four PSSs: Cash, Credit, Leasing and Fee-for-Service (adapted from IEA-PVPS (2003); Krause and Nordström (2004); Worldbank (2008))

	Sales Model		Service Model	
	Cash	Credit	Leasing	Fee for Service
Market potential	Low (< 3 %)	Medium (< 20 %)	Large (< 50 %)	Large (< 70 %)
Ownership	Consumer becomes owner upon payment	Consumer becomes owner through contractual agreement	Service provider is owner during the leasing period, subsequently consumer	Service provider is owner of the system
Initial investment cleared by	Consumer	Financial institute plus down payment by consumer	Service provider and eventually Financial institute	Service provider
Regular installments	No	Yes, to cover the credit	Yes, to cover the rent	Yes, to cover the use of the service
Responsibility of maintenance	Consumer	Consumer and eventually Service provider	Consumer or Service provider	Service provider
Typical maintenance service	No	Often included for a certain time period	At least included during payment period	Included during contract duration
Risk for consumer	High technical risk	Low technical risk	Low technical risk	Very low overall risk
Risk for service provider	Technical risk covered by manufacturer, low financial risk	Technical risk and eventually financial risk	High technical and financial risk	Very high technical and financial risk
Risk for financial institute	n.a.	High financial risk	Medium financial risk for refinancing of the service provider	Medium financial risk for refinancing of the service provider

4.4 Method and data

To evaluate the four PSSs, two stages of data collection are implemented. As a starting point, interviews are conducted with four Indian companies and one German company, all of which are active in the sector. Of the five organisations, two are more commercially focused, whereas the remaining three could be considered more as socially driven businesses. The results from the interviews are used to design a quantitative survey as well as to interpret the results of the survey. One of the key aspects derived from the interviews is that people do have multifaceted views about the meaning of the four PSSs, especially Leasing and Fee-for-Service. This complexity may lead to the confusion of respondents and consequently to misleading results. Therefore, the research team decided not to ask direct questions in the survey such as "Do you prefer Leasing over Fee-for-Service?" but instead to focus on the underlying principles of each PSS. Conjoint analysis is a particularly appealing method for this research as it offers the possibility of evaluating the preferences of respondents (utilities) for the various elements of the PSS.

4.4.1 Conjoint analysis

Based on the work of Luce and Turkey (1964), conjoint analysis (derived from *con*sidering *joint*ly) was first applied in marketing research in the early 1970s (McFadden, 1986; Green and Srinivasan, 1990). The main idea behind this approach is to evaluate the preference for or utility of a stimulus that is composed of several independent attributes, each with a predefined number of levels. Different combinations of carefully selected and defined attribute levels allow the mirroring of reallife decision scenarios to a large extent (see also Fig. 4.2). While conjoint analysis is still mostly used in marketing, other applications are explored, including investment decisions (Riquelme and Rickards, 1992; Lüthi and Prässler, 2011; Loock, 2012) and entrepreneurship (Patzelt and Shepherd, 2009). In the context of this study, entrepreneurs are asked to imagine that they are starting from scratch to set up a business for SHSs in emerging and developing countries. The leading question throughout the survey was as follows: "Which product design represents a reasonable business opportunity for SHS in low-income markets?" The corresponding attributes and levels (Table 4.3) are derived from the PSS presented in Table 4.2, a sample question is presented in Fig. 4.2.

One of the concepts within conjoint analysis is adaptive choice-based conjoint (ACBC). The characteristic of this method is that it adapts subsequent questions during the survey based on answers already given by the respondent (Johnson and Orme, 2007). It does so by recognising attribute levels that are rated to be either absolutely required or not at all required. Consequently, the algorithm focuses on the remaining attributes between these two extremes in remaining questions. This is a more engaging survey experience, as questions ask mostly about combinations of attribute levels that are relevant in the perception of the respondent. From an academic perspective, adaptive choice-based conjoint enables an increase in the accuracy of the responses or reduce the required number of decisions. In other words, the method allows to reduce either the number of questions to each respondent or the sample size without compromising on the level of accuracy. The latter effect is especially relevant for this paper, as the survey sample is comparatively small stemming from a small overall population.

Here you find different sets of product features and income levels. For each one, indicate whether it is a reasonable business model for you or not.

Household income	3000 US$/ year	5000 US$/ year	1000 US$/ year	5000 US$/ year
Your sales service	Advisory service plus on-site installation	Advisory service	Advisory service	No
Your maintenance service	1 year included	5 years included	included during payment period	5 years included
Down payment	100% (cash)	30% & regular installments	10% & regular installments	0%, regular installments
Ownership of the SHS	Private household	Your organisation, after repayment: private household	Your organisation	Private household
	⦿ Reasonable ◯ Won´t work for me	⦿ Reasonable ◯ Won´t work for me	◯ Reasonable ⦿ Won´t work for me	◯ Reasonable ⦿ Won´t work for me

Figure 4.2: Sample question from the ACBC survey

Once the survey is closed, statistical methods such as Hierarchical Bayes (HB) allow the calculation of the utility of each level and for each respondent (Lenk et al., 1996; Johnson and Orme, 2007) To arrive at the overall preferences of all respondents regarding each level, HB estimations calculate in an iterative process the preference of each individual respondent based on the preference of the entire population. This calculation in turn changes the preference of the entire population to some degree, and the individual preferences are calculated again based on these new averages. This process is repeated in 40,000 iterations to curve the stability of the results. In this way, the balance between individual respondents' preferences and sample averages is identified by the amount of variance within the sample (Rossi and Allenby, 2003). Furthermore, the derived preferences or utilities can be an input for simulation methods that estimate the preferences for different combinations of attribute levels. Using them as an input helps, for example, to predict to some extent the market success of new products in comparison with existing products.

Table 4.3: Attributes and levels that are covered in the Conjoint Analysis

Attributes	No.	Levels
Maintenance service	1.1	No
	1.2	Included during contract duration
	1.3	1 year included
	1.4	5 years included
Down payment	2.1	Regular instalments, no down payment
	2.2	10 % down payment & regular instalments
	2.3	30 % down payment & regular instalments
	2.4	100 % (cash)
Sales service	3.1	No
	3.2	Advisory service
	3.3	Advisory service & on-site installation
Ownership of the SHS	4.1	Consumer
	4.2	Your organisation, after repayment: private household
	4.3	Your organisation
Household income	5.1	USD 1 000 per year
	5.2	USD 3 000 per year
	5.3	USD 5 000 per year

4.4.2 Survey sample

To identify experienced decision makers able to provide high quality answers to the survey, a two-step selection process was followed: first, a set of 55 successful companies and organisations in low-income markets that are providing SHSs for households was identified by conducting a comprehensive desk study. The research team focused mainly on international awards and competitions for social businesses as well as on publicly available data from financial institutions and venture capitalists. After identifying the organisations, a list of up to two key decision makers from each organisation was developed. This procedure resulted in a sample of 93 key decision makers in commercially oriented companies that are selling SHSs in one or several emerging and developing countries.

Table 4.4: Description of the sample characteristics

	N	%
Years of professional experience		
up to 5 years	14	45%
5 to 10 years	8	26%
more than 10 years	9	29%
Area of responsibility		
Owner / CEO	18	58%
Top management	7	23%
Other	6	19%
Focus regions (aggregated)[1]		
South America	6	14%
Africa	20	45%
Asia	18	41%
Number of employees		
up to 20	14	45%
20 to 50	10	32%
more than 50	7	23%
Number of SHSs sold		
up to 1 000	14	45%
1 000 to 10 000	8	26%
more than 10 000	9	29%
Refinance of organisations[1]		
Venture Capital / Private Equity	15	48%
Corporate Venture Capital	2	7%
Debt	14	45%
Social Venture Capital	12	39%
Donations	8	26%
Public grants & loans	13	42%

[1] multiple answers possible

Before launching the survey, a pre-test with 9 experts was conducted to validate the measurement and refine the survey. Then, potential respondents are invited individually through personalised messages either by email or via social networks such as LinkedIn. People who did not reply within 2 weeks are contacted twice again. To optimise the accuracy of responses and to limit the impact of self-assessment, it was guaranteed that all information would remain confidential and a promise was made to share with the respondents the study's final results, which would include a personalised feedback document (Huber and Power, 1985). The survey was implemented between September and October 2011.

This process allowed us to gather response data from 31 persons from 27 companies, corresponding to an effective response rate of 33 % in terms of individuals and 49 % in terms of companies in the total population. A total of 15 additional responses are returned incomplete and are thus not taken into account for further analysis. Although the sample size was small, it was possible to cover a relevant share of the global low-income SHSs market. Based on freely available company data, we assigned each company to one or several PSSs as defined in this paper. The comparison of companies in the sample and companies that responded to the survey shows similar patterns (Fig. 4.3). This finding increases the validity of the results by ruling out sample selection and non-response bias. For a detailed sample description see Table 4.4.

Most of the respondents belonged to the top management of their respective organisations. Slightly over half of the respondents had over 5 years of professional experience and ran organisations with more than 20 employees. The companies focused mainly on Africa and Asia, with most respondents focusing on either India (18 %) or Tanzania (16 %). The number of employees, as well as the total number of SHSs sold, indicates that one half of these firms can be considered start-up companies, whereas the other half represents companies with considerable experience with SHSs in low-income markets.

The respondents reveal that their respective organisations rely on private funding, such as venture capital, private equity and bank loans, as well as on financial sources with a broader social scope, such as social venture capital and public grants and loans. One quarter of the organisations relied at least partly on donations as an additional source of revenue. Interestingly, only two organisations benefited from corporate venture capital and are found to be subsidiaries of the same multinational company.

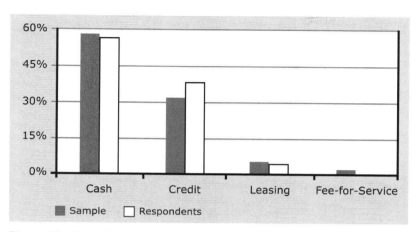

Figure 4.3: Comparison of the applied PSSs of 55 companies in the whole sample and 27 companies of the respondents of the survey (multiple PSSs are possible per company)

4.5 Findings

In the survey, responses regarding a total of 1550 decisions are gathered. This number corresponds to 50 choice tasks per respondent. Other conjoint studies that apply non-adaptive survey techniques already suggest that this number of decisions is sufficient for conducting rigorous analysis (Patzelt et al., 2008; Loock, 2012). Beyond, the advantage of adaptive survey techniques (such as Adaptive Choice Based Conjoint) is the ability to ask more accurate questions that result in a lower number of decisions required; in other words the adaptive nature in our survey selects more relevant items from the overall choice set based on earlier decisions in the course of the survey on the part of the respondents. Based on the survey, HB estimation is able to extract utility values for every level (Table 4.5). Furthermore, the utility values are used to conduct simulation studies on the preference of the sample.

4.5.1 Experimental results of the HB estimation

The utilities of different levels of attributes are zero-centered in order to facilitate the discussion of the results (Fig. 4.4). In general, utility values can be interpreted as follows: A positive utility means that this specific level is reasonable, attractive or useful for the respondent. Correspondingly, a negative utility indicates that this specific level is not attractive or even not reasonable.

Table 4.5: Average utility values of attribute levels and standard deviation

Attribute	Level	Average part-worth	Standard deviation
Maintenance service	No	-0,79743	2,40575
	Included during payment period	0,94105	0,60095
	1 year included	1,24954	0,72625
	5 years included	-1,39317	2,25628
Down payment	0 %, reg. instalments	-2,20251	1,57126
	10 % & reg. instalments	0,12358	1,32459
	30 % & reg. instalments	0,97241	0,98233
	100 % (cash)	1,10652	2,16144
Sales service	No	-1,44892	1,20346
	Advisory service	0,03962	0,93952
	Advisory service plus on-site installation	1,40930	1,08807
Household income	1000 US$/year	0,21392	0,84115
	3000 US$/year	0,34125	0,65560
	5000 USyear	-0,55517	1,14380
Ownership of the SHS	Private household	0,82113	1,04571
	Your organisation, after repayment: private household	0,48728	0,56546
	Your organisation	-1,30841	0,79095

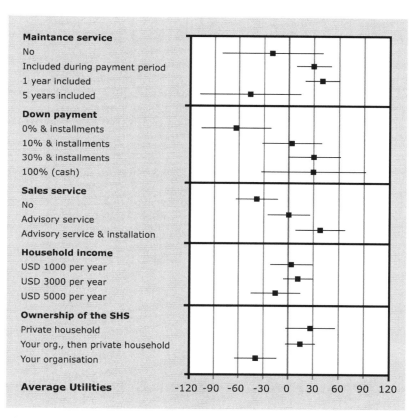

Figure 4.4: Visualisation of the utility values (zero-centered) and standard deviation

Regarding maintenance, it appears that both no maintenance and 5 years of maintenance are not reasonable, but the high standard deviation found shows that there is no general agreement among the respondents. However, the respondents agree that one year of maintenance as well as maintenance services during the payment period are both reasonable from the business perspective of respondents. In addition, the quantitative results fit with qualitative statements by experts who had emphasised – prior to the survey – the link between regular repayments and proper maintenance of the product. One entrepreneur specified that "end-users stop paying their instalments whenever the SHS is not functioning as expected".

Down payments of 0 % are evaluated not to be reasonable from a business perspective. Interestingly, the utilities of a 30 % down payment and a 100 % cash payment are evaluated to be equally reasonable. However, the standard deviation shows that the respondents agree on the first evaluation but have different perceptions regarding the 100 % cash payment. The overall results regarding the payment fit with previous studies that highlight the importance of the end-user's showing real commitment by buying into the SHS as an asset (Vleuten et al., 2007). Our interview partners report that if the consumer does not take responsibility of and make a commitment to the SHS – a challenge in many donor-funded projects – then the system is likely to malfunction after a short period of time. One entrepreneur in India explained during an interview that "to create a sense of responsibility at the consumer side, down payments are absolutely necessary".

The levels related to sales service indicate that SHSs are difficult to sell without any advisory service. In fact, respondents recommended including on-site installations into the services that a organisation should offer. In addition, one entrepreneur clarified that "consumers want to experience the product before they buy it – this is what we are doing to boost the marker".

Regarding ownership of the SHSs, respondents evaluated an intermediate transfer of ownership as well as a transfer of ownership after the completion of repayment, as beneficial. Maintaining the ownership of the SHS rather than handing it over to the customer appears to generate difficulties from the private sector point of view, as one social entrepreneur highlighted: "If the customer knows that at some stage he will own the system, it increases his motivation to pay instalments on time."

Interestingly, household income has the lowest difference in part-worth utility, which means that all three income levels are of interest from a private sector perspective. This finding indicates that household income levels above USD 1000 per year do not represent a barrier for organisations in the context of SHSs and low-income markets.

4.5.2 Business perspective vs. limited financial resources of consumers

The individual preference data are combined to reveal the preference for the four different PSSs. In this way, the share of preference can be observed as a response to the question: "What percentage of entrepreneurs in the field of SHSs would choose this PSS for their business?" and always sums to 100 %. Considering each PSS, we take the levels that result in the maximum individual share of preference within their respective boundaries (e.g., the Cash PSS requires ownership transfer and a 100 % down payment). The described approach results in the "Business preference" scenario (Fig. 4.5; for the specifications, please refer to Table 4.6). If the four "best" PSS compete for the share of preference of the respondents, two effects can be observed. First, almost half of the respondents would prefer a Cash PSS, which is very reasonable from a business perspective. Second, the Fee-for-Service PSS remains marginal, as Credit PSS remains twice as attractive and Leasing PSS approximately three times as attractive.

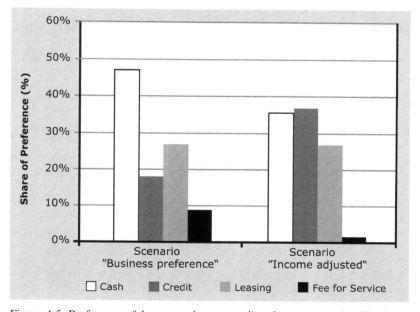

Figure 4.5: Preference of the respondents regarding the two scenarios "Business preference" and "Income adjusted", the sum of the share of preferences equals 100 % (for the definitions, see Table 4.6)

The "Income adjusted" scenario is built on the assumption that more complex PSSs such as Leasing and Fee-for-Service are designed for lower household income levels, which also corre- spond to a reduced availability of cash for down payments. This adjustment results, on the one hand, in an almost equally high preference for both Cash (35 %) and Credit (36 %). On the other hand, the preference for the Leasing PSS did not have relevant changes in preference (26 %), whereas the Fee-for-Service PSS has marginal relevance in the "Income adjusted" scenario.

Table 4.6: Definition of the different attribute levels for each scenario; levels with bold numbers change between both scenarios (see also Table 4.3)

Scenario	PSS	No. of Levels
Business preference	Cash	1.3 2.4 3.3 4.1 **5.2**
	Credit	1.3 2.3 3.3 4.1 5.2
	Leasing	1.3 **2.3** 3.3 4.2 5.2
	Fee-for-Service	1.2 **2.3** 3.3 4.3 5.1
Income adjusted	Cash	1.3 2.4 3.3 4.1 **5.3**
	Credit	1.3 2.3 3.3 4.1 5.2
	Leasing	1.3 **2.2** 3.3 4.2 5.2
	Fee-for-Service	1.2 **2.1** 3.3 4.3 5.1

4.6 Discussion and conclusion

This study demonstrates the relevance of linking products and services in low-income markets. By analysing the case of SHSs, we can show that comparatively expensive products must be combined with services such as advisory services, maintenance, finance and capacity building. The bundling of products and services has been investigated in mature and developed markets, particularly in the context of customised durable goods, such as buildings, trains, airports and hospitals, but has been largely absent from extant research in low-income markets. This explora- tive study is a first attempt to quantify these effects. In this way, this study explores common patterns of success across national borders and beyond regional differences.

In the context of all four PSSs, the Cash variant appears to be an entry point for new companies due to its simplicity and the comparably low capital requirements (Krause and Nordström, 2004), a claim that is supported by current business activities of the sample (Fig. 4.3). As our survey shows, the very same companies evaluate Cash, Credit and Leasing as being almost equally reasonable, which can be interpreted as an indicator for the emergence of more complex PSSs in the near future (Fig. 4.5). Based on the literature and on our interviews, we conclude that this shift towards more complex PSSs results from the motivation of companies to reach significantly more customers beyond the current demand (Worldbank, 2008; Lemaire, 2009). This claim is in line with the traditional PSS literature such as Mont (2002). Pursuing more complex PSSs implies the entrepreneur's taking more risk—either financial risk or asset risk (in case the ownership of the SHS is still with the company). When the entire sector approaches maturity, we expect companies to sell fewer SHSs on a cash basis and pursue Credit and Leasing mechanisms to a significantly greater extent, which results in an inverted-U shape of preferences.

Interestingly, the current trend towards more complex PSSs does not include the Fee-for-Service mechanism. In fact, our survey shows that the low preference of companies for low or zero down payments and the lack of transfer of ownership to the customer are the reasons why Fee-for-Service is not evaluated as a feasible option for purely private companies. This finding resonates with the existing literature in the field of Fee-for-Service models, such as the study of Lemaire (2009), who analysed this mechanism as part of a public–private partnership in Zambia. The joint conclusion is that the risks for purely private companies are too high to implement this mechanism.

Regarding the broader context of diffusion of innovation, this study contributes to a refined understanding of low-income markets. Existing studies focus on either the "high-level" (modelling the entire sector such as Hammond et al. (2007)) or specific case studies (Lemaire, 2009; Mukherji and Jose, 2010). This study aims to bridge the gap between the two camps by drawing from the combined understanding and experience of successful entrepreneurs (micro) across the low-income markets in different countries (macro). This study evaluates how entrepreneurs enable access to energy services in constructive and profitable ways. More specifically – and counterintuitive given prior findings on the market potential of SHSs (Worldbank, 2008) – our results show that entrepreneurs evaluate the Cash PSS to be reasonable and the Fee-for-Service approach not to be feasible without any type of risk sharing or reward, e.g., by public policy or development aid. In addition, this study provides a baseline for future research on "micro" phenomena in the SHSs sector in low-income markets, e.g., through case study research or consumer surveys.

Regarding the method and the sample selection, several implications can be drawn for researchers and policy makers. The method – conjoint analysis – was initially developed for revealing the preferences of consumers regarding new product developments. In line with other studies (Shepherd and Zacharakis, 1999; Patzelt et al., 2008; Patzelt and Shepherd, 2009), this study focuses on entrepreneurs and not the end-users of a product. Conjoint analysis is particularly useful, as it reveals underlying preferences of decision makers for different elements of a PSS and also offers the possibility of conducting simulations with product service combinations that are not explicitly part of the survey. Regarding the sample, a selection process was established to identify the most suitable organisations and decision-makers for the survey. Interestingly, it was not possible to identify any multinational corporation that fits within the selection criteria. Only one multinational company was indirectly involved in the survey, as it owns shares in two small local organisations that match our criteria. One conclusion could be that large multinational companies have not yet managed to overcome the barriers in low-income markets, whereas clever, determined and often socially motivated entrepreneurs are able to set up commercially successful companies in the very same contexts.

4.6.1 Policy implications

The study identified a large gap between supply and demand indicating market failure and thereby scope for policy intervention. Although the study did not explicitly include the policy framework, it has high relevance for the design of public support mechanisms. Both the literature review and the interviews prior to the survey clearly indicate that public support mechanisms such as cash grants can significantly distort a functioning market. One interview partner from India argues that "cash grants add fuel to the fire of corruption". Therefore, more indirect measures, such as cheap refinancing conditions for companies and consumers, appear to be more reasonable for addressing market failure. Beyond financial support that could be provided by favourable refinancing conditions or well-targeted guarantees, capacity building and technical assistance for both policy makers and private sector companies may act as a major lever driving the diffusion of renewable energy technologies. Interestingly, this study also shows that household income levels above USD 1000 per year in principle allow the private sector to deliver SHSs on a commercial basis.[10]

Beyond, three aspects of this study may influence national and international policy makers. First, developing low-income markets with products that are comparatively expensive, such as SHSs, entails designing policy support in a way that focuses not only on the product and its (technical) specifications but also on including services such as advisory pre-sales services, on-site installations and training as well as maintenance and suitable financing services that balance down payments and regular instalments. This study suggests that these aspects are not just important but absolutely essential for successfully developing the private sector in low-income markets. In this way, failures of past public programmes for developing the national off-grid energy sector that are discussed, e.g., by Acker and Kammen (1996) and Vleuten et al. (2007) can at least partly be explained.

Second, the Worldbank (2008) found that the market potential increases when PSSs become more complex. However, this study suggests that more complex PSS mechanisms are not always more suitable from a private sector perspective. In fact, more

---

10   The slight preference expressed for lower household income levels (Fig. 4.4) can be explained by
     the fact that many socially driven organisations participated in the survey.

complex PSSs face additional risks regarding finance and ownership. Therefore, one key conclusion is that the Fee-for-Service PSS could be feasible for the very poor but requires policy intervention to become reasonable for private sector decision makers. In other words, public policy could attempt to correct for a market failure stemming from a mismatch between supply and demand.

Third, because it was not possible to identify large multinational corporations in the SHS space, there is doubt that they are currently able to deliver SHSs to end-users in low-income markets on their own. In fact, entrepreneurs are currently at the forefront to develop the market. One suggestion for policy makers could be to support these organisations in teaming up with local entrepreneurs. Large multinational companies could support local entrepreneurs through their access to the formal (international) market – for example, access to finance and better procurement conditions – but might require a policy "push" to do so.

4.6.2 Future research directions

This study had some limitations that may inspire additional fruitful research. First, a (explorative) quantitative study can only investigate a limited set of aspects. Future research could either refine the aspects covered in this study, such as a differentiation between down payments and collateral, or evaluate other influencing factors that go beyond the scope of this study, such as cultural aspects. Many issues have already been raised and discussed in case studies, and with this study, we would like to encourage other researchers to start quantifying these aspects.

Second, only indirect effects of national and international policies on the preferences of the decision makers have been studied. Public policy mechanisms may have a relevant impact on the entire SHS sector (Sebitosi and Pillay, 2005; Wamukonya, 2007; Vleuten et al., 2007) as well as on the design of the PSS of an individual company, e.g., if public support is only available under certain conditions, such as in the case of Zambia (Lemaire, 2009).

Finally, due to the nature of this nascent sector, the sample size of the survey is comparatively small. Therefore, it is not possible to analyse different sub-groups of the sample. Exploring the variance between different types of companies, regions and decision-makers by quantitative measures will hopefully be possible in the future once more companies enter this market.

5 Conclusions

This thesis contributes to the ongoing debate that surrounds the diffusion of technologies. While many researchers focus on the early stage of technical innovations in emerging countries and beyond (e.g. Tsai et al. (2009); Nemet (2009); Hendry and Harborne (2011)), this thesis explores the diffusion of mature technologies. Thereby, one can abstract from technology risks and reveal the remaining influencing factors. More specifically, this thesis focuses on two technologies for power generation, namely wind farms as an example for grid-connected renewable energy technologies and solar home systems as an example of off-grid renewable energy technologies.

To achieve this, it differentiates between two perspectives on technology diffusion; the aggregated perspective and the micro-perspective (Metcalfe, 1988; MacVaugh and Schiavone, 2010). In the relevant literature, both are applied to the context of mature as well as emerging markets. Focusing on the latter, the thesis attempts to unravel the viewpoint of early adopters from the private sector with regard to different framework conditions and policy mechanisms. The use of the micro-perspective allows the exploration of their underlying preferences and priorities. More specifically, the investigation focuses on project developers and entrepreneurs. Thereby, the point of view of these companies can be interpreted as an early indicator for investment activities. Thus, the micro-perspective firstly confirms existing knowledge generated by other methods and secondly, reveals new insights regarding why, how and when adoption occurs. In light of the initial question of this thesis "what can policy-makers learn from the private sector perspective" (Chapte 1.1), implications not only for policy-makers but also for managers and researchers are drawn.

5.1 Implications for policy-makers

By analysing and revealing perspectives of the private sector regarding renewable energy diffusion in emerging markets, this thesis adds to the debate of how to support the deployment of this type of technology. More specifically, it is intended to help policy-makers take more informed decisions regarding regulations and framework conditions for renewable energy technologies. As policy-makers in the national and international contexts have to allocate limited public resources, trade-offs are unavoidable. In fact, national priorities and development aid agendas, economic and environmental policy as well as social needs (e.g. hospitals and education) and economic obligations (e.g. reliable electricity supply) all compete for limited public funding (Zerriffi and Wilson, 2010; Bambawale et al., 2011). However, if public resources have been made available for renewable energy technologies, this thesis may facilitate effective and efficient processes such as drafting regulations and assuring their proper implementation.

In line with the implications for policy-makers discussed below, this thesis implicitly ties into a vital debate going on in the realm of international climate change negotiations. Since the Conference of the Parties in 2009 in Copenhagen (COP 15), the idea of creating an international fund, called the Green Climate Fund[11], is being discussed. Three questions regarding the design of the fund have been brought to the fore; those of fund mobilisation, fund governance and fund disbursement (Bird et al., 2011). This thesis contributes to the third question on how to disburse public funding that effectively and efficiently leverages private investments: First, it clearifies the relevance for country specific and well targeted guarantees for private investments in grid-connected renewable energy infrastructure and second, it highlights the need for developing suitable mechanisms that facilitate refinancing of companies in the off-grid renewable energy space.

Beyond the detailed recommendations discussed below, a vital conclusion of the research is the need for increased knowledge exchange between the public and the private sector. This is necessary in order to increase both, the knowledge of public decision-makers regarding the current potentials and limitations (risks) of renewable energy technologies, and the knowledge of private decision-makers regarding constraints and opportunities in the public decision-making process. The following recommendations are given for each empirical context (grid-connected and off-grid).

5.1.1 Grid-connected renewable energy technologies

Wind farms are chosen as a case study for the empirical context of grid-connected renewable energy technologies. Drawing from Chapters 2 and 3, several recommendations for policy-makers are made. Based on the findings of this thesis a step-by-step approach is developed for policy-makers in emerging and developing countries that aim to increase the share of grid-connected renewable energy power generation through private sector investments. It consists of five elements:

---

11   The Green Climate Funds's prerogative is to raise and distribute 100 billion USD per annum, by channeling both public funding and private investments into adaptation and mitigation measures in emerging and developing countries. In this way, the fund will leverage private funding for renewable energy technologies and thereby help these countries benefit from technology leapfrogging (Soete, 1985; Huberty and Zysman, 2010; Mathews et al., 2011; Sierra, 2011).

1. *Understand the opportunity by revealing the realisable market potential*: This is done through analytical triangulation of available natural resources, technical opportunities such as available area, and technical constraints on power generation technology or grid infrastructure. The technical market potential for each renewable energy technology is thus revealed (see also Chapter 2 in IEA (2008)). This first step was mentioned during the interviews but is not part of this thesis, which assumes that natural resources and the necessary infrastructure are available in abundance.

2. *Reveal rationale for private sector investments*: This step forms the core of this thesis. By first revealing and understanding the rationale and the risks for private sector investments, policy-makers can decide if and how they address these risks. Depending on the country the results can be very different as one project developer in the context of policy support for wind farms stated: "Please do not use the watering can principle but well-targeted guarantees for each country" (WIG5). This conclusion is based on the follwing experience of project developers in Chile and Argentina: While interview partners find it difficult to manage the approval process in Chile, international credit agencies and investors seem to be ready to invest (WIG5, WIG8). In Argentina interview partners explained that it is somewhat easier to get approvals. However, due to the recent expropriation in the oil sector, international investors are currently reluctant to invest (WIG4, WIG5).

Based on the findings of this thesis presented and discussed in Chapter 2.3, Table 2.3 and Chapter 3.4, a checklist is developed that covers identified risks (Table 5.1). In addition, the findings indicate that public and private sector decision-makers[12] have a shared interest in reducing investment risks. In fact, refinancing conditions are a key influencing factor for overall costs of grid-connected renewable energy technologies. Thus, a higher perceived risk e.g. regarding the electricity tariff for renewable energy (associated with support mechanisms like the quota system) results in significantly higher overall costs for electricity generated. This finding from de Jager and Rathmann (2008) and Butler and Neuhoff (2008) in the industrialised country context hold also true for emerging markets.

3. *Create and show commitment to private and public stakeholders*: Setting ambitious targets for renewable energy diffusion is a strong signal for both involved government agencies as well as national and international private sector decision-makers. Ambitious targets that are defined by high-level politicians can help to align the often diverging interests between different public authorities as mentioned above. However, beyond setting ambitious targets, it is crucial to develop institutional and technical capacities (see also Chapter 2.3). At the institutional level, it is important to discuss and potentially provide training to all involved stakeholders, such as public authorities at the national, district and local level, grid operators, the local private sector (man-

---

12   for the private sector perspective see also Table 3.2

ufacturers, transport and construction companies), local public and private banks and universities. At the technical level, it might be helpful to properly define the interface between power plant and grid infrastructure (standard setting) and to discuss long-term plans for the extension or upgrade of the existing grid.

Table 5.1: Selection of risks and potential measures for wind farms

Risk	Potential measures
Electricity tariff risk[1]	Implement guaranteed price floors
	Provide government guarantees
Unclear definition of support mechanism	Capacity building for public stakeholders
	Consult with private sector
Unclear interface with grid infrastructure	Define and implement clear criteria
	Capacity building for grid operators
	Consult with private sector
Legal risks[2]	Strengthen the legal system
	Provide government guarantees
Risk of policy change	Communicate upcoming changes in a transparent manner
Gap between local content requirements and local expertise	Evaluate current capabilities of local industry
	Support for joint ventures
	Consult with private sector
	Reduce local content requirements
Insufficient supply of capital	Provide guarantees (currency or inflation risk)
	Provide public co-investments
	Capacity building for local investors and local credit agencies
	Consult with private sector

[1] e.g. by quota system or due to low financial stability of the off-taker
[2] e.g. land use rights, expropriation

4. *Emphasise on implementation of regulations and commitments*: Beyond the definition of the support mechanism, additional guarantees and the interface with the grid infrastructure, it is vital to focus on proper implementation in order to have the intended impact on private investments. Although this step seems to be common sense, the findings of this thesis reveal it as a major challenge in emerging markets (see also Chapter 2.3 and Chapter 3.4.2). During the workshop, a wind turbine manufacturer clearified: "You never know how regulations are implemented – although they are approved by the government!" (WWG6). Well-implemented regulations reduce uncertainty for project developers and investors and consequently also their respective return expectations. In the case of infrastructure investments in a broader sense, this

effect has been quantified by Henisz (2002) who provides a two-century-long historical analysis of 100 countries. The qualitative data analysed in this thesis suggests an
equally strong effect in emerging and developing countries (see also Chapter 2.3). This
includes measures to make government decisions and decision criteria as transparent
as possible to private sector decision-makers, thereby creating and increasing trust in
government authorities.

5. *Adapt the policy framework in a predictable way*: In many cases, the framework
for private sector investments in grid-connected renewable energy plants needs to be
adjusted in order to reflect cost reductions through technical improvements and new
emerging risks and opportunities. In the context of the EU Lüthi and Wüstenhagen
(2012) quantified the effect of policy change for project developers and found that
unexpected policy changes in the last 5 years result in significantly higher return expectations. In the case of emerging markets private sector decision-makers suggest
implementing these changes in a transparent and predictable way in order to allow local and international companies to adjust their operations accordingly (see also Chapter
3.4.1). For policy-makers, predictability is especially relevant for firstly, the development of a domestic industry in a nascent market (Lewis and Wiser, 2007) and secondly,
the transformation of the energy systems from a formative phase into one characterised
by positive feedbacks (Jacobsson and Bergek, 2004).

5.1.2 Off-grid renewable energy technologies

By selecting solar home systems as a case study, this thesis offers insights into the
broader category of off-grid renewable energy technologies. Drawing from Chapter
2 and Chapter 4, several recommendations for policy-makers can be made. From a
broader perspective, private sector decision-makers highlight the importance of public commitment and support at the national and international level. Although is not
feasible to connect all rural areas to the national grid – even in the long run - it
seems as if policy-makers underestimate the opportunities that off-grid power supply
offers.

Past efforts to develop the sector are often well-intentioned but are based on donations
or donor-driven projects (Wamukonya, 2007; Vleuten et al., 2007). These programmes
show mixed results as one social investor underlined during an interview: "Before one
of our companies starts selling its solar home systems in a new region, we must first
clean up broken solar home systems from previous government programmes and donations" (OIE4). Investigating the relation between company and customer can provide
a partial explanation why so many goverment programmes failed (see also Chapter
4). In fact, commercially oriented companies that sell solar home system in emerging
markets focus on maintenance and financial services as well as customer commite-

ment (e.g. through down payment) and education (see also Chapter 2.4). In line with Mondal et al. (2010), the raised issues with government support programmes indicate that the entire sector is shifting from national and international development aid towards a market-based approach. More specifically, facilitating market development through well-targeted measures that do not disturb the nascent but emerging market for off-grid applications like SHS should be the new focus of policy-makers. Potential risks and measures based on this explorative research design are presented in Table 5.2.

Table 5.2: Selection of risks and potential measures for solar home systems

Risk	Potential measure
Market distortion by low quality products	Develop quality standards or labels Strengthen customer awareness
Market distortions by grants	Use grants only for well targeted projects[1] Implement support mechanisms that do not distort the market[2]
Market distortions by fossil fuel subsidies	Allow for an equal playing field[3]
Lack of skilled employees	Develop training and education jointly with universities and industry (associations)
Lack of capital supply for companies	Implement PPP funds for refinancing Develop opportunities for public co-financing Provide grants for the due diligence process especially for investments below 3 mio US$ Capacity building of local private investors and credit agencies
Lack of capital supply for customers	Provide refinancing for companies Strengthen the microfinance sector
Lack of maintenance	Adjust public tenders accordingly Train local technicians Educate the customer

[1] e.g. for school buildings and public buildings
[2] e.g. by measures that help the whole sector (capacity bilding, reduced import tax for PV panels)
[3] e.g. by subsidised vouchers that customers can use either for paying back for a solar home system or for buying fossil fuels

Beyond specific measures, and in line with the argument to develop a local market for off-grid power, policy-makers need to focus their attention more on investments and less on grants. Regarding grants, one interview partner in India explained that the government asks for a certain level of product quality in order to provide 30 % grant for each solar home system sold which seems to be a good idea. However, the public quality check of the products lacks transparency which leads to the conclusion that "grants add fuel to the fire of corruption" (OII2). This thesis finds that beyond a few exeptions that are discussed at the beginning of Chapter 2.4, grants are percieved as a risk and are associated with a lack of transparency and non-sustainable market development by the interview partners and workshop participants.

Regarding investments, the qualitative data derived from the interviews in India and Tanzania reveal that most entrepreneurs face the challenge of raising funds in order to provide financial services such as a 2-year repayment period to their customers (see also Chapter 2.4). The gap becomes even more visible as international whole-salers deliver mostly on a prepayment basis. Interestingly, this gap exists not because the refinancing needs are so high but because the refinancing needs are too low. In fact, the need for refinancing of the Tanzanian interview partners ranges from 50 000 USD to 500 000 USD. These comparably low amounts of money cannot be provided by current finance mechanisms. Public investors such as donor agencies look for investment opportunities beyond 10 million USD due to high transaction costs. Private banks perceive high risks regarding country, technology and company and also high costs for due diligence. So far, this market failure is addressed to a very limited extend by specialised (international) social investors that are in principle able to provide debt and equity up to 1 million USD. Although this funding gap seems to be highly relevant for market development, it is not jet addressed by both practicionners and academia.

Managing the transition from a publicly funded pilot project to public private part-nership funds to the privately funded scale-up of business by local banks is a key for long-term success. Currently interview partners in both countries are trying to develop a revolving fund for their customers in order to address this barrier for growth and market development. However, one workshop participant raised the issue that, for him, it is currently easier to obtain a grant from a social investor (due to lower transaction costs) than to raise private and public investments for his projects in Africa (OWG2). As a next step and in order to go beyond pilot projects and focus on scaling up, the interview partners and workshop participants suggest developing a revolving private or public private partnership fund that offers (low interest) loans to local companies.

5.2 Implications for managers

Beyond the question of what policy-makers can and should do from a private sector perspective, several recommendations are applicable to private sector decision-makers themselves. Depending on the context, four approaches could make a valuable contribution to market development in emerging and developing countries.

Firstly, it is recommendable to support and nurture local capacities within emerging markets. This could include developing training measures for existing and future local employees (e.g.in cooperation with universities and other companies) to explicitly engage in local procurement, educating local technology suppliers and local investors and credit agencies on renewable energy investments (see also Chapter 2.3 and 2.4).

Secondly, building "unconventional" partnerships between different public and private stakeholders to share risks might significantly increase the rate of success for upcoming projects. This could potentially tackle the challenge of working capital for the case of off-grid power supply or risk exposure for the case of grid-connected power plants. As an example one interview partner is currently developing wind farms and large photovoltaic power plants in order to provide electricity for the mining industry. Selling electricity directly to large industrial consumers allows to significantly reduce many risks summarised in Table 5.1 above.

Thirdly, for the case of off-grid power supply, new combinations of products and services might prove to be effective (Chaurey et al., 2012). In line with the Worldbank (2008), one can conclude that a huge demand for rural electrification services is not yet exploited. Based on a quantitative survey among commercially oriented companies that sell solar home systems in these markets, a trend towards more complex product service mechanism (from Cash to Credit to Leasing) could be observed (see also Chapter 4). However, it has also emerged that contrary to the application in industrialised countries (see for example Mont and Lindhqvist (2003); Hypko et al. (2010)), the Fee-for-Service mechanism is not yet suited to the private sector in low-income markets.

Finally, all these activities provide sound arguments that may convince policy-makers to further develop favourable framework conditions. However, it is crucial for private sector decision-makers to engage in dialogue within the industry and with public policy-makers.

5.3 Implications for researchers

Among the first researchers that investigate in the diffusion of innovation is Griliches (1957) who finds that adoption of hybrid corn seeds in the US depends on the specific value of applying the seed in a given region. Later, Mansfield (1989) deduces that riskiness of new technology and the amount of required investments influence adoption. Other researchers focus especially on the influence of institutions and regulation on diffusion of technologies. By conducting an extensive literature review on technology diffusion, Jaffe et al. (2003) conclude that an exhaustive ranking of policy instruments for the diffusion of innovation cannot be based on theory alone. They further clearified that limited empirical evidence for policy options in the field of environmentally beneficial technologies is available.

Regarding the question of which policies and institutional settings are most suitable for driving the diffusion of power generation technologies in a given region, scholars have not come to a final conclusion yet. By looking at the example of wind farm policies in Germany and UK, Wong (2005) concludes that both regulation (Germany) and deregulation (UK) can drive the diffusion of wind energy if adjusted appropriately to the context. Others (e.g. Lipp (2007); Butler and Neuhoff (2008)) argue that deregulation (quota systems) results in lower deployment and higher costs compared to regulation (feed-in-tariff), as investment risks are higher in more competitive environments. In line with the latter argument, this thesis finds that reducing risks for early adopters is even more relevant in emerging markets as the investment risks are considerably higher compared to countries like Germany and the UK. The quantiative survey reveals that feed-in-tariffs are the most attractive support mechanism (see also Chapter 3.4.2). Other specific risks and measures that are identified to be relevant are discussed in the previous chapter on policy implications (Table 5.1).

In line with the rationale for this thesis, Popp et al. (2010) and Blackman and Kildegaard (2010) specify that the context of developing and emerging countries is key for research on the diffusion of technologies as the results can potentially be very different from the context of industrialised countries. In fact, the current academic debate on energy policies focuses on the question which policy mechanisms are most effective and efficient (e.g., Buen (2006); Sovacool (2010); Nemet (2009)). However, as these studies are almost exclusively conducted in the context of industrialised countries, they assume that full implementation follows the policy design process. This assumption does not hold in the case of emerging and developing countries. In fact, this thesis finds that a proper implementation is a key influencing factor for early adopters to start developing renewable energy projects in an emerging market (see also Chapter 2.3 and Table 2.3).

More specifically, it also seems to matter who is driving the implementation. Weber et al. (2009) closely linked the process of diffusion with the adoption of policies in emerging economies by local decision-makers. More specifically, they analysed the global diffusion of stock markets between 1980 and 2005 and find that donor-driven initiatives in emerging countries result in a less vibrant national stock market while initiatives driven by local stakeholders show positive effects. However, aligning the often divergent interests of stakeholders seems to be a challenge. This conclusion is based on the case of off-grid energy systems (see also Chapter 2.4). The shared commitment for deploying solar home systems and other off-grid technologies by policy-makers, NGOs and companies at the district and local level has the potential to tremendously push technology diffusion (see also Chapter 2, Table 2.3) sometimes even before regulations are in place (Popp et al., 2011).

Different interests of stakeholders as well as lack of communication between the public and the private sector can lead to significant but neccessary policy changes[13]. By analysing infrastructure investments in over 100 countries over a period of two-centuries Henisz (2002) could show that a high risk of policy change has a negative effect in investments in infrastructure. The findings of this thesis support this argument (Table 3.2) and add to it by clarifying how to avoid the negative effects of policy change. In fact, the process of implementing necessary policy changes should follow a step-by-step approach and be communicated in a transparent manner so as to strengthen the trust of private sector decision-makers (see also Chapter 5.1).

In order to gain the support of relevant stakeholders and design and implement effective and efficient framework conditions for investments, capacity building emerges as one key instrument to facilitate technology diffusion. Thereby this thesis confirms existing studies that focus on policy (Vleuten et al., 2007; IEA, 2010), company (Bell et al., 1984; Prahalad, 2012), non-govermnetal organisations (Chesbrough et al., 2006; Perez-Aleman and Sandilands, 2008), university (Schiller, 2006) or end-user levels (Dewald and Truffer, 2011; Schillebeeckx et al., 2012). Drawing from the barriers for technology diffusion that emerge in this thesis (Chapter 2, Table 2.3), it is critical to extend capacity building activities to local public and private investors and credit agencies. One interview partner in Tanzania highlighted "The African Development Bank recently conducted a workshop with us [entrepreneurs] and local credit agencies - however, they [local credit agencies] seem to be not interested in the rural energy business"(OIT1). This is surprising to some extend as one might assume that credit agencies and investors are looking for new investment opportunities themselves.

The use of the micro-perspective of individual decision-makers in this thesis allows the

---

13 In Chapter 3.4.1 the example of several Eastern European countries that required several years including several policy changes until the market could develop in a sustainable way is briefly discussed.

exploration of why and when adoption occurs in resource-limited settings and consequently helps to draw more specific recommendations and conclusions than the aggregated perspective. Thereby the thesis confirms existing knowledge (such as the need for predictability) and reveals "blind spots" that could not be revealed by applying the macro-perspective of technology diffusion (such as lack of implementation of regulations and lack of refinancing opportunities for entrepreneurs).

No research is without limitations, and this thesis acknowledges its own. Firstly, it is solely focused on analysing the perspective of the private sector. While it is very relevant to understand the rationale and perception of private sector decision-makers, other aspects such as welfare considerations can and should influence the decisions of policy-makers. Secondly, it discusses early adopters within emerging markets, which is but one element of the value chain. Although this limitation could partly be addressed by including decision-makers from international wholesalers and manufacturers during both workshops, further studies might be required. Thirdly, in both cases the validity of the quantitative results is limited to the given scenario of the survey. Using data from the survey in order to draw conclusions for different applications such as village grids or much larger wind farms is only possible to a limited extent. For the case of solar home systems, there are other important influencing factors that go beyond the scope of this thesis. Many of these influencing factors have already been raised and discussed in case studies (see for example Mair and Schoen (2007); Lemaire (2009); Anderson et al. (2010)) and this thesis would like to encourage other researchers to start quantifying these aspects. For the case on wind power, it was not possible to analyse actual investment flows or expected rates of return. Therefore, we cannot predict investment flows that will result from policy recommendations that are made in this thesis. Finally, due to the empirical setting of emerging markets and in line with methodological challenges raised by George et al. (2012), the sample size for both qualitative and quantitative data is comparatively small although it covers several regions. It will be interesting to explore the variance between different types of companies, regions and decision-makers by quantitative measures.

Future research on the diffusion of technology in emerging markets could take several paths. One path would be to conduct longitudinal studies that map how the percpetion of key decision-makers change over time. Another path would be to map the perspective of all involved public and private stakeholders in one or several countries. A third approach could be large quantitative studies. Hopefully, this will be possible in the near future once more companies in the renewable energy space set up their business activities in emerging and developing countries.

Bibliography

Abdullah, S., Mariel, P., 2010. Choice experiment study on the willingness to pay to improve electricity services. Energy Policy 38, 4570–4581.

Acker, R., Kammen, D., 1996. The quiet (energy) revolution - Analysing the dissemination of photovoltaic power systems in Kenya. Energy Policy 24 (1), 81–111.

AGECC, 2010. Energy for a Sustainable Future - Report and Recommendations. The Secretary-General's Advisory Group on Energy and Climate Change.

Aitken, M., 2010. Why we still don't understand the social aspects of wind power: A critique of key assumptions within the literature. Energy Policy 38 (4), 1834 – 1841.

Allard, G., Martinez, C., Williams, C., 2012. Political instability, pro-business market reforms and their impacts on national systems of innovation. Research Policy 41, 638–651.

Anderson, J., Markides, C., Kupp, M., 2010. The Last Frontier: Market Creation in Conflict Zones, Deep Rural Areas, and Urban Slums. California Management Review 52 (4), 6–28.

Athreye, S., Cantwell, J., 2007. Creating competition? Globalisation and the emergence of new technology producers. Research Policy 36, 209–226.

Bachman, J., O'Malley, P., 1984. Yea-saying, nay-saying, and going to extremes: Black White differences in response style. The Public Opinion Quarterly 48 (2), 491–509.

Bambawale, M., D'Agostino, A., Sovacool, B., 2011. Realizing rural electrification in Southeast Asia: Lessons from Laos. Energy for Sustainable Development 15, 41–48.

Bazilian, M., Nussbaumer, P., Eibs-Singer, C., Brew-Hammond, A., Modi, V., Sovacool, B., Ramana, V., Aqrawi, P.-K., 2012. Improving Access to Modern Energy Services: Insights from Case Studies. The Electricity Journal 25 (1), 93–114.

Bell, M., Ross-Larson, B., Westphal, L., 1984. Assessing the Performance of infant industries. Journal of Development Economics 16, 101–128.

Bergek, A., Jacobsson, S., 2010. Are tradable green certificates a cost-efficient policy driving technical change or a rent-generating machine? Lessons from Sweden 2003-2008. Energy Policy 38, 1255–1271.

Bewley, T., 2002. Interviews as a valid empirical tool in economics. Journal of Socio-Economics 31, 343–353.

Bird, N., Brown, J., Schalatek, L., 2011. Design challenges for the Green Climate Fund. Heinrich Böll Stiftung North America.

Blackman, A., Kildegaard, A., 2010. Clean Technological Change in Developing-Country Industrial Clusters: Mexican Leather Tanning. Environmental Economics and Policy Studies 12 (3), 115–132.

Brew-Hammond, A., 2010. Energy access in Africa: Challenges ahead. Energy Policy 38, 2291–2301.

Buen, J., 2006. Danish and Norwegian wind industry: The relationship between policy instruments, innovation and diffusion. Energy Policy 34, 3887–3897.

Butler, L., Neuhoff, K., 2008. Comparison of feed-in tariff, quota and auction mechanisms to support wind power development. Renewable Energy 33, 1854–1867.

Byrne, R., 2011. Micro-energy systems in low-income countries: learning to articulate the solar home system niche in Tanzania. In: Schäfer, M., Kebir, N., Philipp, D. (Eds.), Micro Perspectives for Decentralized Energy Supply - Proceedings of the International Conference. TU Berlin, pp. 207–219.

Casillas, C., Kammen, D., 2011. The delivery of low-cost, low-carbon rural energy services. Energy Policy 39, 4520–4528.

Chaurey, A., Kandpal, T., 2010. Assessment and evaluation of PV based decentralized rural electrification: An overview. Renewable and Sustainable Energy Reviews 14 (8), 2266–2278.

Chaurey, A., Krithika, P., Palit, D., Rakesh, S., Sovacool, B., 2012. New partnerships and business models for facilitating energy access. Energy Policy 47, 48–55.

Chesbrough, H., Aherne, S., Finn, M., Guarrez, S., 2006. Business Models for Technology in the Developing World: The Role of Non Governmental Organizations. California Management Review 48 (3), 48–61.

Chircu, A., Mahajan, V., 2009. Perspective: Revisiting the Digital Divide: An Analysis of Mobile Technology Depth and Service Bredth in BRIC Countries. Journal of Product Innovation Management 26, 455–466.

Cohen, S., Neira, L., May 2003. Measuring preference for product benefits across countries: overcoming scale usage bias with maximum difference scaling. In: ESO-MAR 2003 Latin America Conference Proceedings. Amsterdam.

Cohen, S., Orme, B., 2004. What's your preference? Marketing Research, 32–37.

Couch, A., Keniston, K., March 1960. Yeasayers and naysayers: agreeing response set as a personality variable. Journal of Abnormal and Social Psychology 60, 151–174.

Couture, T., Gagnon, Y., 2010. An analysis of feed-in tariff remuneration models: Implications for renewable energy investment. Energy Policy 28, 955–965.

de Jager, D., Rathmann, M., 2008. Policy instrument design to reduce financing costs in renewable energy technology projects. IEA Implementing Agreement on Renewable Energy Technology Deployment.

del Rio González, P., 2005. Analysing the Factors Influencing Clean Technology Adoption: A Study of the Spanish Pulp and Paper Industry. Business Strategy and the Environment 14, 20–37.

Dewald, U., Truffer, B., 2011. Market Formation in Technological Innovation Systems - Diffusion of Photovoltaic Applications in Germany. Industry and Innovation 13 (3), 285–300.

Dinica, V., 2006. Support systems for the diffusion of renewable energy technologies - an investor perspective. Energy Policy 34, 461–480.

Dinica, V., 2008. Initiating a sustained diffusion of wind power: The role of public-private partnerships in Spain. Energy Policy 36, 3562–3571.

Eisenhardt, K., 1989. Building theories from Case Study Research. The Academy of Management Review 14 (4), 532–550.

Enzensberger, N., Wietschel, M., Rentz, O., 2002. Policy instruments fostering wind energy projects–a multi-perspective evaluation approach. Energy Policy 30, 793–801.

Farizo-Alvarez, B., Hanley, N., 2002. Using conjoint analysis to quantify public preferences over the environmental impacts of wind farms. an example from spain. Energy
Policy 30, 107–116.

Finn, A., Louviere, J., 1992. Determining the Appropriate Response to Evidence of
Public Concern: The Case of Food Safety. Journal of Public Policy and Marketing
11 (2), 12–25.

Flyvbjerg, B., 2006. Five Misunderstandings About Case-Study Research. Qualitative
Inquiry 12 (2), 219–245.

Garud, R., Karnoe, P., 2003. Bricolage versus breakthrough: distributed and embedded
agency in technology entrepreneurship. Research Policy 32, 277–300.

George, G., McGahan, A., Prabhu, G., 2012. Innovation for Inclusive Growth: Towards a Theoretical Framework and a Research Agenda. Journal of Management
Studies 49, 661–683.

George, G., Prabhu, G., 2003. Developmental financial institutions as technology
policy instruments: implications for innovation and entrepreneurship in emerging
economies. Research Policy 32, 89–108.

Geroski, P., 2000. Models of technology diffusion. Research Policy 29, 603–625.

Green, P., Srinivasan, V., 1990. Conjoint Analysis in Marketing: New Developments
With Implications for Research and Practice. Journal of Marketing, 3–19.

Griliches, Z., 1957. Hybrid corn: An exploration in the economics of technical change.
Econometrica 48, 501–522.

Gross, R., Blyth, W., Heptonstall, P., 2009. Risks, revenues and investment in electricity generation: Why policy needs to look beyond costs. Energy Economics 32 (4),
796–804.

Guerin, T., 2009. An Assessment and Ranking of Barriers to Doing Environmental
Business with China. Business Strategy and the Environment 18, 380–396.

Gustavsson, M., 2007. Educational benefits from solar technology - Access to solar
electric services and changes in children's study routines, experiences from eastern
province Zambia. Energy Policy 35, 1292–1299.

Gustavsson, M., Ellegard, A., 2004. The impact of solar home systems on rural livelihoods. Experiences from the Nyimba Energy Service Company in Zambia. Renewable Energy 29, 1059–1072.

Haas, R., Eichhammer, W., Huber, C., Langniss, O., Lorenzoni, A., Madlener, R., Menanteau, P., Morthorst, P., Martins, A., Oniszk, A., Schleich, J., Smith, A., Vass, Z., Verbruggen, A., 2004. How to promote renewable energy systems successfully and effectively. Energy Policy 32, 833–839.

Hall, B., 2005. Innovation and Diffusion.

Hammond, A., Kramer, W., Katz, R., Tran, J., Walker, C., 2007. The Next 4 Billion - Market Size and Business Strategy at the Base of the Pyramid. World Resources Institute, International Finance Corporation, Washington DC.

Hargadon, A., 2010. Technology policy and global warming: Why new innovation models are needed. Research Policy 39, 1024–1026.

Hendry, C., Harborne, P., 2011. Changing the view of wind power development: More than "bricolage". Research Policy 40 (5), 778–789.

Henisz, W., 2002. The institutional environment for infrastructure investment. Industrial and Corporate Change 11 (2), 355–389.

Holt, D., 2011. Where Are They Now? Tracking the Longitudinal Evolution of Environmental Businesses From the 1990s. Business Strategy and the Environment 20, 238–250.

Huber, G., Power, D., 1985. Retrospective reports of strategic-level managers: guidelines for increasing their accuracy. Strategic Management Journal 6, 171–180.

Huberty, M., Zysman, J., 2010. An energy system transformation: Framing research choices for the climate challenge. Research Policy 39, 1027–1029.

Hypko, P., Tilebein, M., Gleich, R., 2010. Clarifying the concept of performance-based contracting in manufacturing industries. Journal of Service Management 21 (5), 625–655.

IEA, 2008. Deploying Renewables – Principles for Effective Policies. International Energy Agency, Paris.

IEA, 2010. Comparative Study on rural Electrification Policies in Emerging Economies. International Energy Agency.

IEA, 2011. Deploying Renewables – Best and Future Policy Practice. International Energy Agency, Paris.

IEA, UNDP, UNIDO, 2010. Energy Poverty - How to make modern energy access universal? International Energy Agency, United Nations Development Programme, United Nations Industrial Development Organization.

IEA-PVPS, 2003. Summary of Models for the Implementation of Photovoltaic Solar Home Systems in Developing Countries. IEA Implementing Agreement on Photovoltaic Power Systems (Report IEA PVPS T9-02: 2003).

IRENA, June 2012. Renewable Energy Jobs and Access. International Renewable Energy Agency.

Jacobson, A., Kammen, D., 2007. Engineering, institutions, and the public interest: Evaluating product quality in the Kenyan solar photovoltaics industry. Energy Policy 35, 2960–2968.

Jacobsson, S., Bergek, A., 2004. Transforming the energy sector: the evolution of technological systems in renewable energy technology. Industrial and Corporate Change 13 (5), 815–849.

Jacquet, J. B., 2012. Landowner attitudes toward natural gas and wind farm development in northern pennsylvania. Energy Policy 50 (0), 677 – 688.

Jaffe, A., Newell, R., Stavins, R., 2003. Handbook of Environmental Economics. Vol. 1. Elsevier Science Publishers, Ch. Technological Change and the Environment, pp. 462–516.

Jefferson, M., 2008. Accelerating the transition to sustainable energy systems. Energy Policy 36 (11), 4116 – 4125.

Johnson, R., Orme, B., 2007. A New Approach to Adaptive CBC. Sawtooth Software.

Karekezi, S., Kithyoma, W., 2002. Renewable energy strategies for rural Africa: is a PV-led renewable energy strategy the right approach for providing modern energy to the rural poor of sub-Saharan Africa? Energy Policy 30, 1071–1086.

Kaundinya, D., Balachandra, P., Ravindranath, N., 2009. Grid-connected versus stand-alone energy systems for decentralized power - A review of literature. Renewable and Sustainable Energy Reviews 13, 2041–2050.

Kemp, R., Oltra, V., 2011. Research Insights and Challenges on Eco-Innovation Dynamics. Industry and Innovation 18 (3), 249–253.

Ketlogetswe, C., Mothudi, T., 2009. Solar home systems in Botswana - Opportunities and constraints. Renewable and Sustainable Energy Reviews 13, 1675–1678.

Khanna, T., Palepu, K., 1997. Why Focused Strategies May Be Wrong for Emerging Markets. Harvard Business Review, 41–51.

Kleimeier, S., Versteeg, R., 2010. Project finance as a driver of economic growth in low-income countries. Review of Financial Economics 19 (2), 49–59.

Komatsu, S., Kaneko, S., Shresta, R., Ghosh, P., 2011. Nonincome factors behind the purchase decision of solar home systems in rural Bangladesh. Energy for Sustainable Development 15 (3), 284–292.

Komendantova, N., Patt, A., Barras, L., Battaglini, A., 2009. Perception of risk in renewable energy projects: The case of concentrated solar power in North Africa. Energy Policy 40, 103–109.

Krause, M., Nordström, S. (Eds.), 2004. Solar Photovoltaics in Africa - Experiences with financing and delivery models. United Nations Development Programme and Global Environment Facility, New York, NY.

Kristinsson, K., Rao, R., 2008. Interactive Learning or Technology Transfer as a Way to Catch-Up? Analysing the Wind Energy Industry in Denmark and India. Industry and Innovation 15 (3), 297–320.

Lee, G., 2009. Understanding the timing of "fast-second" entry and the relevance of capabilities in invention vs. commercialization. Research Policy 38, 86–95.

Leitner, A., Wehrmeyer, W., France, C., 2010. The impact of regulation and policy on radical eco-innovation. Management Research Review 33, 1022–1041.

Lemaire, X., 2009. Fee-for-service companies for rural electrification with photovoltaic systems: The case of Zambia. Energy for Sustainable Development 13, 18–23.

Lenk, P., DeSarbo, W., Green, P., Young, M., 1996. Hierachical Bayes Conjoint Analysis: Recovery of Part worth Heterogeneity from Reduced Experimental Designs. Marketing Science 15 (2), 173–191.

Levin, T., Thomas, V., 2012. Least-cost network evaluation of centralized and decentralized contributions to global electrification. Energy Policy 41, 286–302.

Lewis, J., Wiser, R., 2007. Fostering a renewable energy technology industry: An international comparison of wind industry policy support mechanisms. Energy Policy 35, 1844–1857.

Lindner, K., 2011. Quality Issues In The Market Based Dissemination Of Solar Home Systems. In: Schäfer, M., Kebir, N., Philipp, D. (Eds.), Micro Perspectives for Decentralized Energy Supply - Proceedings of the International Conference. TU Berlin, pp. 220–228.

Lipp, J., 2007. Lessons for effective renewable electricity policy from Denmark, Germany and the United Kingdom. Energy Policy 35, 5481–5495.

London, T., Anupindi, R., Sehth, S., 2010. Creating mutual value: Lessons learned from ventures serving base of pyramid producers. Journal of Business Research 63, 582–594.

Loock, M., 2012. Going beyond best technology and lowest price: on renewable energy investors preference for service-driven business models. Energy Policy 40, 21–27.

Luce, D., Turkey, J., 1964. Simultaneous Conjoint Measurement: A New Type of Fundamental Measurement. Journal of Mathematical Psychology 1, 1–27.

Lund, P., 2009. Effects of energy policies on industry expansion in renewable energy. Renewable Energy 34, 53–64.

Lüthi, S., Prässler, T., 2011. Analyzing policy support instruments and regulatory risk factors for wind energy deployment – a developer's perspective. Energy Policy 39 (9), 4876–4892.

Lüthi, S., Wüstenhagen, R., 2012. The price of policy risk – Empirical insights from choice experiments with European photovoltaic project developers. Energy Economics 34 (4), 1001–1011.

MacVaugh, J., Schiavone, F., 2010. Limits to the diffusion of innovation - A literature review and integrative model. Europan Journal of Innovation Management 13 (2), 197–221.

Mair, J., Schoen, O., 2007. Successful social entrepreneurial business models in the context of developing economies - An explorative study. International Journal of Emerging Markets 2 (1), 54–68.

Mansfield, E., 1989. Industrial robots in Japan and the USA. Research Policy 18, 183–192.

Markhard, J., Truffer, B., 2006. Innovation processes in large technical systems: Market liberalization as a driver for radical change? Research Policy 35, 609–625.

Martinot, E., Chaurey, A., Lew, D., Moreira, J. R., Wamukonya, N., 2002. Renewable Energy Markets in Developing Countries. Annu. Rev. Energy Environ. 27, 309–348.

Mathews, J., Hu, M.-C., Wu, C.-Y., 2011. East-Follower Industrial Dynamics: The Case of Taiwan's Emergent Solar Photovoltaic Industry. Industry and Innovation 18 (2), 177–202.

Mathews, J., Kidney, S., Mallon, K., Hughes, M., 2010. Mobilizing private finance to drive an energy industrial revolution. Energy Policy 38, 3263–3265.

McEachern, M., Hanson, S., 2008. Socio-geographic perception in the diffusion of innovation: Solar energy technology in Sri Lanka. Energy Policy 36, 2578–2590.

McFadden, D., 1986. The Choice Theory Approach to Market Research. Marketing Science 5 (4), 275–297.

Metcalfe, J., 1988. Technical change and economic theory. Ch. The diffusion of innovations: an interpretative survey.

Mondal, A. H., Kamp, L., Paachova, N., 2010. Drivers, barriers, and strategies for implementation of renewable energy technologies in rural areas in Bangladesh - An innovation system analysis. Energy Policy 38, 4626–4634.

Mont, O., 2002. Clarifying the concept of product-service system. Journal of Cleaner Production 12, 237–245.

Mont, O., Lindhqvist, T., 2003. The role of public policy in advancement of product service systems. Journal of Cleaner Production 11, 905–914.

Moran-Ellis, J., Alexander, V., Cronin, A., Dickinson, M., Fielding, J., Sleney, J., Thomas, H., 2006. Triangulation and integration: processes, claims and implications. Qualitative Research 6 (1), 45–59.

Mostafaeipour, A., October 2010. Productivity and development issues of global wind turbine industry. Renewable and Sustainable Energy Reviews 14 (3), 1048–1058.

Mowery, D., Nelson, R., Martin, B., 2010. Technology policy and global warming: Why new policy models are needed (or why putting new wine in old bottles won't work). Research Policy 39, 1011–1023.

Mukherji, S., Jose, P., 2010. Selco: Harnessing Sunlight to Create Livelihood. Harvard Business Review.

Nakata, C., Weidner, K., 2012. Enhancing New Product Adoption at the Base of the Pyramid: A Contextualized Model. Journal of Product Innovation Management 29 (1), 21–32.

Ndzibah, E., 2010. Diffusion of solar technology in developing countries - Focus group study in Ghana. Management of Environmental Quality: An International Journal 21 (6), 773–784.

Nemet, G., 2009. Demand-pull, technology-push, and government-led incentives for non-incremental technical change. Research Policy 38, 700–709.

Noethlichs, S., 2011. Wind Energy in Bulgaria – from gold rush to standstill or from 0 MW to 14000 MW and back again. In: European Wind Energy Association (EWEA) Annual Conference 2011.

Obeng, G., Akuffo, F., Braimah, I., Evers, H.-D., Mensah, E., 2008. Impact of solar photovoltaic lighting on indoor air smoke in off-grid rural Ghana. Energy for Sustainable Development 12 (1), 55–61.

Ondraczek, J., 2012. The Sun Rises in the East (of Africa): A Comparison of the Development and Status of the Solar Energy Markets in Kenya and Tanzania, Working Paper FNU-197.

Orme, B., 2002. CVA/HB Technical Paper. Sawtooth Software.

Patzelt, H., Shepherd, D., January 2009. Strategic Entrepreneurship at Universities: Academic Entrepreneurs' Assessment of Policy Programs. Entrepreneurship Theory and Practice 33 (1), 319–340.

Patzelt, H., Shepherd, D., Deeds, D., Bradley, S., 2008. Financial slack and venture managers' decision to seek a new alliance. Journal of Business Venturing 23, 465–481.

Perez-Aleman, P., Sandilands, M., 2008. Building Value at the Top and the Bottom of the Global Suppy Chain: MNC-NGO Partnerships. California Management Review 51 (1), 24–49.

Popp, D., 2010. Innovation and Climate Policy. NBER Working Paper No. 15673.

Popp, D., Hafner, T., Johnstone, N., 2011. Environmental policy vs. public pressure: Innovation and diffusion of alternative bleaching technologies in the pulp industry. Research Policy 40 (9), 1253–1268.

Popp, D., Newell, R., Jaffe, A., 2010. Handbook of the Economics of Innovations. Vol. 2. Burlington Academic Press, Ch. Energy, The Environment, and Technological Change, pp. 873–938.

Porter, M., 1990. The Competitive Advantage of Nations. FreePress, NewYork.

Prahalad, C., 2012. Bottom of the Pyramid as a Source of Breakthrough Innovations. Journal of Product Innovation Management 29 (1), 6–12.

Ranjit, L., O'Sullivan, K., 2002. A Sourcebook for Poverty Reduction Strategies. Vol. 2. World Bank, Washington, DC.

Rao, P., Miller, J., Wang, Y., Byrne, J., 2009. Energy-microfinance intervention for below poverty line households in India. Energy Policy 37, 1694–1712.

Rebane, K., Braham, B., 2011. Knowledge and adoption of solar home systems in rural Nicaragua. Energy Policy 39, 3064–3075.

REN21, 2012. Renewables 2012 Global Status Report. Renewable energy policy network for the 21st century.

Ringel, M., 2006. Fostering the use of renewable energies in the European Union: the race between feed-in tariffs and green certificates. Renewable Energy 31, 1–17.

Riquelme, H., Rickards, T., 1992. Hybrid Conjoint Analysis: An Estimation Probe in new Venture Decisions. Journal of Business Venturing 7, 505–518.

Rogers, 1995. Diffusion of innovations, 4th Edition. Free Press, New York.

Rossi, P., Allenby, G., 2003. Bayesian Statistics and Marketing. Marketing Science 22 (3), 304–328.

Sarkar, J., 1998. Technological Diffusion: Alternative Theories and Historical Evidence. Journal of Economic Surveys 12 (2), 131–176.

Schillebeeckx, S., Parikh, P., Bansal, R., George, G., 2012. An integrated framework for rural electrification: Adopting a user-centric approach to business model development. Energy Policy 48, 687–697.

Schiller, D., 2006. Nascent Innovation Systems in Developing Countries: University Responses to Regional Needs in Thailand. Industry and Innovation 13 (4), 481–504.

Schilling, M., Esmundo, M., 2009. Technology S-curves in renewable energy alternatives: Analysis and implications for industry and government. Energy Policy 37, 1767–1781.

Schleicher-Trappeser, R., 2012. How renewables will change electricity markets in the next five years. Energy Policy 48, 64–75.

Schrader, C., Freimann, J., Seuring, S., 2011. Business Strategy at the Base of the Pyramid. Business Strategy and the Environment 21 (5), 281–293.

Seawright, J., Gerring, J., June 2008. Case Selection Techniques in Case Study Research - A Menu of Qualitative and Quantiative Options. Political Research Quarterly 61 (2), 294–308.

Sebitosi, A., Pillay, P., 2005. Energy services in sub-Saharan Africa: how conductive is the environment? Energy Policy 33, 2044–2051.

Seelos, C., Mair, J., 2007. Profitable Business Models and Market Creation in the Context of Deep Poverty: A Strategic View. Academy of Management Perspectives 21 (4), 49–63.

Shepherd, D., Zacharakis, A., 1999. Conjoint analysis: a new methodological approach for researching the decision policies of venture capitalists. Venture Capital 1 (3), 197–217.

Sierra, K., 2011. The Green Climate Fund: Options for Mobilizing the Private Sector. Brookings Institution.

Skidmore, D., 1997. Encouraging Sustainable Business Practice Through Managing Policy Instruments Dynamically. Business Strategy and the Environment 6, 163–167.

Söderholm, P., 2008. Harmonization of renewable electricity feed-in laws: A comment. Energy Policy 36, 946–953.

Soete, L., 1985. International Diffusion of Technology, Industrial Development and Technological Leapfrogging. World Development 13 (3), 409–422.

Sovacool, B., 2009. Rejecting renewables: The socio-technical impediments to renewable electricity in the United States. Energy Policy 37, 4500–4513.

Sovacool, B., 2010. A comparative analysis of renewable electricity support mechanisms for Southeast Asia. Energy 35, 1779–1793.

Sovacool, B., D'Agostino, A., Bambawale, M., 2011. The socio-technical barriers to Solar Home Systems (SHS) in Papua New Guinea: "Choosing pigs, prostitutes, and poker chips over panels". Energy Policy 39, 1532–1542.

Späth, P., Rohracher, H., 2010. Energy regions: The transformative power of regional discourses on socio-technical futures. Research Policy 39, 449–458.

Stenzel, T., Frenzel, A., 2008. Regulating technological change – The strategic reactions of utility companies towards subsidy policies in German, Spanish and UK electricity markets. Energy Policy 36, 2645–2657.

Tsai, Y., Lin, J., Kurekova, L., 2009. Innovative Research and Development and optimal investment under uncertainty in high-tech industries: An implication for emerging economies. Research Policy 38, 1388–1395.

Tukker, A., Tischner, U., 2006. Product-services as a research field: past, present and future. Reflections from a decade of research. Journal of Cleaner Production 14, 1552–1556.

Urmee, T., Harries, D., Schlapfer, A., 2009. Issues related to rural electrification using renewable energy in developing countries of Asia and Pacific. Renewable Energy 34, 354–357.

Vleuten, F., Stam, N., Plas, R., 2007. Putting solar home system programmes into perspective: What lessons are relevant? Energy Policy 35, 1439–1451.

Wamukonya, N., 2007. Solar home system electrification as a viable technology option for Africa's development. Energy Policy 35, 6–14.

Wamukonya, N., Davis, M., 2001. Socio-economic impacts of rural electrification in

Namibia: comparisons between grid, solar and unelectrified households. Energy for Sustainable Development 5 (3), 5–13.

Watson, J., Sauter, R., 2011. Sustainable innovation through leapfrogging: a review of the evidence. International Journal Technology and Globalisation 5 (3/4), 170–189.

Weber, K., Davis, G., Lounsbury, M., 2009. Policy as Myth and Ceremony? The Global Spread of Stock Exchanges, 1980-2005. Academy of Management Journal 52 (6), 1319–1347.

Wijayatunga, P., Attalage, R., 2005. Socio-economic impact of solar home systems in rural Sri Lanka: a case-study. Energy for Sustainable Development 9 (2), 5–9.

Wiser, R., Pickle, S., 1998. Financing investments in renewable energy: the impacts of policy design. Renewable and Sustainable Energy Reviews 2, 361–389.

Wong, S.-F., 2005. Obliging Institutions and Industry Evolution: A Comparative Study of the German and UK Wind Energy Industries. Industry and Innovation 12 (1), 117–145.

Worldbank, 2008. Operational Guidance for World Bank Group Staff: Designing Sustainable Off-Grid Rural Electrification Projects: Principles and Practices. World Bank Energy and Mining Sector Board.

Yin, R., 2011. Qualitative Research from Start to Finish. The Guilford Press, New York.

Zerriffi, H., Wilson, E., 2010. Leapfrogging over development? Promotion rural renewalbes for climate change mitigation. Energy Policy 38, 1689–1700.

Appendix

Appendix for Chapter 2

Table 5.3: Key questions during the interviews

Case	Questions
Wind Farms	What is your role and responsability within the organisation?
	What barriers do you face in each of your current country offices outside of the EU?
	How do you decide whether to enter a new market or not?
	What is your experience with different types of investors? (debt, equity, public, private, national, international)
	If the german government would like to spend 100 million Euros to deploy wind farms outside the EU, what should they do?
Solar Home Systems	What is your role and responsability within the organisation?
	What is your core business?
	What barriers to market deployment do you face?
	What is your experience with different financing sources? (debt, equity, national, international, public, private)
	What should policy makers do to support the off-grid sector?

Table 5.4: Qualitative data sources for case 1 on wind farms

Code	Organisation	Title	Expertise
WIG1	Project developer	Director International Projects	Turkey, South America
WIG2	Project developer	Assistant of CEO	Hungary, Turkey, Bulgaria, Romania
WIG3	Project developer	Owner	Taiwan, South America, Turkey
WIG4	Project developer	Regional director	Argentina, Bulgaria
WIG5	Project developer	Regional director	Argentina, Chile
WIG6	Project developer	Managing director	Romania, Ukraine
WIG7	Project developer	Managing director	South Africa, Eastern Europe
WIG8	Project developer	Managing director	Southern America, Eastern Europe
WIE1	Finance expert	General manager	Southern Africa
WIE2	Policy think tank	Head of department	BRICS
WIE3	Industry association	Policy analyst	Eastern Europe
WIE4	Finance expert	Head of department	Project finance
WWG1	Public investor	Investment manager	Financing wind farms
WWG2	Assurance company	Head of department	Risk management
WWG3	Industry association (assurance)	Technical expert	Risk management
WWG4	Project developer	Project manager	Eastern Europe
WWG5	Project developer	Financial expert	South America
WWG6	Manufacturer	Director	New markets
WWE1	Industry association	Vice president	Marocco, Poland
WWE2	Industry association	President	Romania
WWE3	Industry association	Policy analyst	Emerging markets
WWE4	Consultant	Senior consultant	Policy, technology
WWE5	Consultant	Junior consultant	Public policy
WWE6	Consultant	Manager	Law
WWE7	Consultant	Owner	Eastern Europe

Table 5.5: Qualitative data sources for case 2 on solar home systems

Code	Organisation	Title	Expertise
OII1	Manufacturer and local company	Vice President	Product development, India
OII2	Local company	Director, owner	India
OII3	Local company	Director	India
OII4	Local company	Director, owner	India
OIT1	Local company	Director, owner	Tanzania
OIT2	Local company	Regional director	Tanzania
OIT3	Local company	Director, owner	Tanzania
OIT4	Local company	Manager	New market development, Africa
OIT5	Local company	Director, owner	Tanzania
OIE1	NGO	Department manager	Rural electrification, Tanzania
OIE2	Industry association	Executive secretary	Tanzania
OIE3	Public agency	Director general	Finance and policy
OIE4	Financial expert	COO	Specialised Fund
OWG1	Wholesaler	Project manger	Off-grid in Africa
OWG2	Wholesaler	Director, owner	Senegal
OWG3	Wholesaler	Head of department	Off-grid in Africa
OWG4	Wholesaler	Head of department	Off-grid power in tourism sector
OWE1	Consultant	Director, owner	Microfinance
OWE2	Consultant	Head of department	Monitoring and evaluation, Bangladesh
OWE3	Public agency	Project manager	South Africa, public research projects
OWE4	Consultant	Owner	Public finance

Appendix for Chapter 3

For the Maximum Difference Scaling survey a decision scenario for project developers was developed based on the results of the 11 interviews. A sample question from the survey is presented in Figure 5.1.

Imagine you have to decide whether your company develops a 30 MW wind park in an emerging economy or not. The following conditions are given:

- *The political stability is acceptable.*

- *So far, only few wind parks have got connected to the grid.*

- *Government authorities have agreed on ambitious long-term targets and financial support for wind energy.*

- *First estimations regarding the internal rate of return (IRR) are promising.*

Imagine you are deciding whether to develop a 30MW greenfield wind park in an emerging country or not.

Which is the **most attractive** and which is the **least attractive** feature?

Most attractive		Least attractive
○	Trained local technicians in the field of wind energy are available	○
○	Grid access and dispatch are regulated but not guaranteed	●
●	"one-stop-shop" for all necessary approvals	○
○	Quota-System with Tradable Green Certificates	○
○	The off-taker has a good credit rating	○

Figure 5.1: Sample question from the survey

Finanzmärkte und Klimawandel

Herausgegeben von Dirk Schiereck und Paschen von Flotow

Band 1 Christian Babl / Paschen von Flotow / Dirk Schiereck (Hrsg.): Projektrisiken und Finanzierungsstrukturen bei Investitionen in erneuerbare Energien. 2011.

Band 2 Christoph Ettenhuber: Financing Corporate Growth in the Renewable Energy Industry. 2013.

Band 3 Anette von Ahsen / Robert Fraunhoffer / Dirk Schiereck (Hrsg.): Wellenbrecher auf dem Weg zur Energiewende? Zur Attraktivität von Energiespeicherung, nachhaltiger Erzeugung und Verbrauchersteuerung. 2013.

Band 4 Christian Friebe: Diffusion of Renewable Energy Technologies. Private Sector Perspectives on Emerging Markets. 2014.

www.peterlang.com